D0927112

Biodiversity

GRACE LIBRARY CARLOW UNIVERSITY
PITTSBURGH PA 15213

Biodiversity

Christian Lévêque
Centre National de la Recherche Scientifique, France

Jean-Claude Mounolou
Université d'Orsay, France

QH
541.15
B56
L4812
2003

John Wiley & Sons, Ltd

CATALOGUED

First published in French as *Biodiversité, Dynamique biologique et conservation* © 2001 Dunod, Paris

Translated into English by Vivien Reuter.

This work has been published with the help of the French Ministère de la Culture-Centre national du livre

English language translation copyright © 2003 by John Wiley & Sons Ltd,
The Atrium, Southern Gate, Chichester,
West Sussex PO19 8SQ, England
Telephone (+44) 1243 779777

Email (for orders and customer service enquiries): cs-bookswilcy.co.uk
Visit our Home Page on www.wileyeurope.com or www.wiley.com

All Rights Reserved. No part of this publication may be reproduced, stored in a retrieval system or transmitted in any form or by any means, electronic, mechanical, photocopying, recording, scanning or otherwise, except under the terms of the Copyright, Designs and Patents Act 1988 or under the terms of a licence issued by the Copyright Licensing Agency Ltd, 90 Tottenham Court Road, London W1T 4LP, UK, without the permission in writing of the Publisher. Requests to the Publisher should be addressed to the Permissions Department, John Wiley & Sons Ltd, The Atrium, Southern Gate, Chichester, West Sussex PO19 8SQ, England, or emailed to permreqwiley.co.uk, or faxed to (+44) 1243 770571.

This publication is designed to provide accurate and authoritative information in regard to the subject matter covered. It is sold on the understanding that the Publisher is not engaged in rendering professional services. If professional advice or other expert assistance is required, the services of a competent professional should be sought.

Other Wiley Editorial Offices

John Wiley & Sons Inc., 111 River Street, Hoboken, NJ07030, USA

Jossey-Bass, 989 Market Street, San Francisco, CA 94103-1741, USA

Wiley-VCH Verlag GmbH, Boschstr. 12, D-69469 Weinheim, Germany

John Wiley & Sons Australia Ltd, 33 Park Road, Milton, Queensland 4064, Australia

John Wiley & Sons (Asia) Pte Ltd, 2 Clementi Loop #02-01, Jin Xing Distripark, Singapore 129809

John Wiley & Sons Canada Ltd, 22 Worcester Road, Etobicoke, Ontario, Canada M9W 1L1

British Library Cataloguing in Publication Data

A catalogue record for this book is available from the British Library

ISBN 0 470 84956 8 hardback
0 470 84957 6 paperback

Typeset in 10.5 on 13pt in Times New Roman
by Kolam Information Services Pvt. Ltd, Pondicherry, India
Printed and bound in Great Britain by TJ International Ltd, Padstow, Cornwall
This book is printed on acid-free paper responsibly manufactured from sustainable forestry in which at least two trees are planted for each one used for paper production.

Contents

Biodiversity   Christian Lévêque and Jean-Claude Mounolou
© 2004 John Wiley & Sons, Ltd   ISBN 0 470 84956 8 (Hbk)   ISBN 0 470 84957 6 (pbk)

Introduction

In less than a century, our perception of nature and the living world has changed profoundly. Ample evidence of this shift is found in social behaviour and in schoolbooks.

In the early 20th century, the world population was primarily rural, and survival was its top priority. Predators and crop pests abounded, and harvests were uncertain. In Europe and the tropics (this was still the colonial era), humans were seriously threatened by diseases. The natural and animal worlds were often perceived as hostile. Thus, up until the mid-20th century, schoolbooks classified animals as 'harmful' or 'useful.' There was a national economic stake in destroying 'harmful' species in order to foster agricultural development. As one French schoolbook taught children, in preparation for adult life: 'Almost all insects are harmful and must be fiercely combated.'

This attitude was entirely legitimate at the time. In daily life, humankind suffered unbearable attacks, especially in the agricultural domain (crop pests) and to health (malaria, for example). Given this psychological context, it is not surprising that people sometimes went too far. Attitudes towards birds of prey, for example, manifest an ignorance of nature and how it operates, a psychotic relationship to wild species, and an exaltation of human supremacy over nature. 'Birds of prey, bandits! All such birds are bandits worse than bandits. If they are only the slightest bit harmful, that is reason enough for me to eradicate them' (extract from *The French Hunter*, 1924).

What were scientists doing during this time? They were collecting, inventorying, and drawing up lists of animal and plant species in different regions, in the tradition of natural history cabinets. They were also actively involved in the national battle against crop pests.

After the Second World War, behavioural patterns slowly began to change: urbanization and industrialization progressed. Many citizens became distanced from the rural world. The development of insecticides

Biodiversity   Christian Lévêque and Jean-Claude Mounolou
© 2004 John Wiley & Sons, Ltd   ISBN 0 470 84956 8 (Hbk)   ISBN 0 470 84957 6 (pbk)

encouraged people to believe that sooner or later it would be possible to control harmful insects such as the Colorado beetle, the locust, the cockchafer, as well as mosquitoes. DDT, later decried for causing ecological damage, was initially hailed as a miracle product that would finally liberate humans from certain natural constraints and give them better control over agricultural production. This was also the time of the 'Green Revolution', of intensive farming based on high-yield crops, but at the same time requiring intensive use of fertilizers and insecticides.

In the early 1970s, the epithets 'harmful' and 'useful', as applied to animals, disappeared from schoolbooks. The whole idea behind such classifications was called into question. The 1960s also saw the beginnings of the science of ecology. Henceforth, knowledge was no longer structured around species, but rather based upon the functioning of natural systems and the relationships between different animal and plant species that constitute 'ecosystems'.

By the 1980s, the human populations of the western hemisphere had come to dominate most of their predators (or so-called predators . . .) and had acquired technologies for controlled and intensive farming. They had finally achieved their ends, as defined by the prevailing mentality at the beginning of the century; i.e. they were well on the way to overcoming natural constraints. And yet the situation today is far from idyllic, and a new perception of nature has taken shape in Western society. Under the pressure of conservation movements (who represent conservationism, not ecological science), there is a growing sense of guilt over the destruction of species that was encouraged in preceding decades. The large NGOs for nature conservation have played an important role in sharpening public awareness for the disappearance of charismatic species, especially mammals and birds. On the other hand, citizens see nature as a place of repose, of recreation and resources. They want 'nature' to be attractive (beautiful landscapes), welcoming (not too many mosquitoes) and full of life (animals and plants to look at). Intensive farming, with its immoderate use of pesticides and fertilisers and destruction of hedges and trees, has been increasingly called into question for its ecological consequences.

People began to talk about the environment in the 1970s. The farmer, once considered the mainstay of the national economy and gardener of 'natural' spaces, was marginalized and accused, sometimes rightly, of destroying landscapes, fauna and flora. At the same time, in the tropical world, scientists and conservationists are concerned by the large-scale destruction of forests regarded as hotspots of living nature. Humans stand accused: they are held accountable for the erosion of biological diversity on the face of the Earth as a result of their uncontrolled activities. The term 'biodiversity' was invented to qualify this impact of human activities upon natural environments and the species that inhabit them. Biodiversity became a global concern, culminating in the Rio Conference on Sustainable Development in 1992. In the process, the debate shifted from the scientific to the political arena.

One thing leads to another; it is urgently necessary to take action to preserve biological diversity, if we do not want to be the agents and witnesses of new mass extinctions. Planning and realizing the appropriate measures requires both scientific knowledge and political will. Conventions are being signed, reserves created, and efforts are underway to implement a somewhat simplistic application of the principle of sustainable development. Some people are driven by ethical considerations: we must preserve the world as we inherited it for the benefit of our children. Others need to be convinced by more pragmatic reasoning; biological diversity is presented as an economic resource of the first order – as a reservoir of genes and molecules useful to agriculture, pharmaceutics and industry. The commercialization of the living world creates new economic prospects with biotechnology and patents on living things. Given the stakes, it is logical to take measures to conserve a source of wealth that has so far been only partially turned to profit.

Within the scientific community, research directions and foci of interest are diversifying. Genetic sequencing and molecular biology are affording ever deeper insight into the living world. The old question of the origin of life has resurfaced, but accompanied this time by knowledge and tools that may deliver concrete answers. Biotechnology offers new prospects for using the living world through genetic engineering of organisms. The economic stakes are huge, but new ethical and scientific questions arise as to the limits of genetically modified organisms (GMO) and the conditions for their use.

Thanks to advances in genetics and new knowledge derived from palaeontology, the great adventure of evolution has once again captured the public interest. At the same time, the inventory of species is continuing with new methods and tools (ecology, physiology, molecular biology, databases, etc.). For a long time, life was considered to be constrained by its physical and chemical environment; however, recent studies in ecology and palaeontology have shown that life actually contributes extensively towards modifying and shaping its environment. The living world plays an active role in the dynamics of the major biogeochemical cycles that are partly responsible for climate states and changes.

In practical terms, the conservation of biological diversity raises both technical and social questions. To implement the principles of sustainable development, the central concern of all conservation policy, it is necessary to find compromises between species protection and development.

In less than a century, the behaviour of western societies towards nature has changed profoundly. They have gradually moved away from their initial impulse to control a hostile natural world towards a more respectful approach to life, seeking a balance that meets the demands of humanity without destroying the diversity of the living world. Nature is still seen as useful, but there is now also concern for protecting nature so as to improve future prospects of exploiting resources yet to be discovered. This change in attitude springs from motivations that are both ethical, aesthetic, commercial and ecological. All these aspects work together, such that it is difficult to evaluate their respective import.

At the same time, we are experiencing an exhilarating period in science. Never before has our knowledge about the living world advanced at such a pace. On the one hand, we are extending the frontiers for the infinitesimally small; on the other, we are developing tools for exploring our planet in its entirety and searching for traces of life in the Universe. Seen through the prism of biological diversity, the debate over humans versus nature and the origins of humanity acquires a new dimension. In the search for solutions to the future of biological diversity, of which humans are a component, it is important to transcend the barriers of academic disciplines and relate the social with the natural sciences. The future of biodiversity cannot be reduced to a technical problem; it depends upon the economic and social choices facing societies in coming decades. In some sense, it depends upon the attitude of each and every citizen.

The aim of this book is to illuminate some perspectives of this issue by giving the reader an overview of current knowledge about the diversity of the living world and the various problems entailed in its conservation and sustainable use.

1 Brief History of a Concept: Why be Concerned by Biological Diversity?

The term 'biodiversity' is perceived differently, depending upon the sociological group involved. Taxonomists, economists, agronomists and sociologists each have their own partial view of the concept. Biologists tend to define biodiversity as the diversity of all living beings. Farmers are interested in exploiting the manifold potential deriving from variations over soils, territories and regions. Industry sees a reservoir of genes useful in biotechnology or a set of exploitable biological resources (timber, fish, etc.). As for the general public: its main concern is with landscapes and charismatic species threatened by extinction. All these points of view are admissible, since the concept of biodiversity effectively refers to a variety of different concerns. Moreover, these different approaches are not independent of one another; they implicitly pursue the same objective, namely the conservation of natural environments and the species which they harbour.

Biodiversity emerged as an environmental issue in the early 1980s, culminating in the Conference on Sustainable Development held in Rio in 1992. Towards the end of the 20th century, humankind grew conscious of its unprecedented impact upon natural environments and the danger of exhausting biological resources. At the same time, biological diversity was recognised as an essential parameter, in particular for the agro-alimentary and pharmaceutical industries. This raised ethical questions about the conservation of biological diversity and patenting of living beings.

Biodiversity   Christian Lévêque and Jean-Claude Mounolou
© 2004 John Wiley & Sons, Ltd   ISBN 0 470 84956 8 (Hbk)   ISBN 0 470 84957 6 (pbk)

Thus, biodiversity became a framework for considering and discussing the whole range of questions raised by human relationships with other species and natural environments – a kind of 'mediator', as it were, between ecological systems and social systems. Independently of this new role, biodiversity remains one of the major concerns relative to global environment.

1.1   What does 'Biodiversity' Refer to?

The term 'biodiversity' – a contraction of biological diversity – was introduced in the mid-1980s by naturalists who were worried about the rapid destruction of natural environments such as tropical rainforests and demanded that society take measures to protect this heritage. The term was adopted by the political world and popularized by the media during the debates leading up to the ratification of the Convention on Biological Diversity.

The expression actually covers a number of essentially different approaches, orientated around four major issues.

- Due to technological progress and the need to occupy new spaces to meet the demands of a rapidly growing population, humankind is impacting natural environments and the diversity of living resources to an unprecedented degree. The questions raised by this tendency vary considerably, as do the possible responses, depending upon the behaviour and choices of particular societies in their approach to economic development. Ultimately, it is a matter of implementing strategies for conservation so as to preserve the natural patrimony as the heritage of future generations (Figure 1.1).

- To understand the causes and conditions that have led to the diversity of the living world as we know it today, we need a new perspective on evolutionary processes. What are the biological mechanisms that explain species diversity? What are the interactions between changes in the biophysical environment and in the phenomena of speciation? Our knowledge of such matters remains fragmentary. While it is still important to continue with the process of making an inventory of species that was initiated by Linnaeus in the 18th century, we must also exploit modern methodological advances to penetrate the world

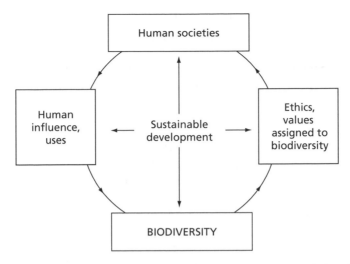

Figure 1.1 Interactions between human societies and biological diversity

of the infinitely minute and the molecular mechanisms involved in the
diversification of life.

- Advances in ecology are also redefining our approach to biological
  diversity as the product of dynamic interactions among different
  levels of integration within the living world. We are now aware that
  the living world acts upon and modifies its physical/chemical environment. The functional processes of ecosystems, such as the flows of
  matter and energy, are subject to the twofold influence of both
  physical/chemical and biological dynamics. This realization constitutes a major paradigm shift, challenging the customary tendency to
  consider only the influence of the physical/chemical context upon the
  dynamics of the living world, to the exclusion of other interactions.
  This integrated approach leads to new concepts such as functional
  ecology and biocomplexity (see box).

- Finally, biodiversity is seen as 'useful' nature – the set of species and
  genes that humankind uses for its own profit, whether they are
  derived from natural surroundings or through domestication. In
  this context, biodiversity becomes a natural form of capital, subject
  to the regulatory forces of the market and a potential source of
  considerable profit to countries possessing genetic resources. The

economic valuation of biodiversity also provides powerful arguments for the cause of natural conservationists.

Biological diversity, biodiversity, biocomplexity

Indiscriminate usage of the word 'biodiversity' may generate indifference or even aversion to the term. We propose a more specific application that focuses on questions of interactions between humankind and nature. Traditionally, the term biodiversity has been used with regard to the depletion of the living world as a result of human activities, or activities undertaken for its protection and conservation – whether through creation of protected areas or by modifying human behaviour with respect to development (the concept of sustainable development).

Here, the term biodiversity will be used to refer to the whole range of activities traditionally connected with inventorying and studying living resources.

The term biological complexity, or biocomplexity, belongs to the new scientific vocabulary of biodiversity. Biocomplexity is the result of functional interactions between biological entities, at all levels of organization, and their biological, chemical, physical and social environments. It involves all types of organisms from microbes to humans, all kinds of environments from polar spheres to temperate forests to agricultural regions, and all human activities affecting these organisms and environments. Biocomplexity is characterized by non-linear, chaotic dynamics and interactions on different spatiotemporal scales. Integrating social and economic factors, it deepens our understanding of the living system in its entirety, rather than in bits and pieces.

1.2   The Origins of the Convention on Biological Diversity and What is at Stake

The ratification of the Convention on Biological Diversity by a majority of nations marks a new chapter in our consciousness of the risks pertaining to the erosion of biological diversity. Today, the issue is seen as an environmental concern of global dimensions demanding urgent solutions. To a certain extent, the approach to this question resembles that to climate change. Both discussions converge upon a similar statement –

humankind is exerting a collective impact of unprecedented magnitude upon the Earth as a whole. Nothing will remain exempt from its effect!

The Preamble to the Convention addresses the role of biological diversity in the biosphere, humankind's responsibility for the depletion of biological diversity, the lack of knowledge needed to undertake appropriate measures for its conservation, the preference for preserving ecosystems and natural habitats rather than resorting to *ex situ* measures. At the same time, the Preamble also acknowledges that economic and social development are priorities for the developing countries and that nations have sovereign rights over the exploitation and conservation of their biological resources. Altogether, the Convention represents a political compromise among diverse concerns and communities of interest.

1.2.1 The 'conservationists'

People have long been concerned by the extinction or near-disappearance of species such as the aurochs and bison in Europe, the dodo on the island of Mauritius, the emperor penguin of Antarctica and the American migrant pigeon. The depletion of these emblematic species is largely the result of intensive hunting by humans.

In recent decades, the magnitude of human impact upon natural environments has attained unprecedented dimensions. Significant population growth, utilization of previously untouched territories and increasing efficiency of technological means of exploitation are given as the major causes. Biodiversity loss no longer means only the extinction of isolated species but rather the modification of entire ecosystems, with all their floristic and faunistic components. Towards the end of the 1970s, naturalists drew attention to the rapid destruction of certain environments such as tropical rainforests. The American zoologist E.O. Wilson declared that humankind was the cause for species extinctions on a par with the mass extinctions of the past. Others have gone so far as to prophesy the end of life on Earth, with humankind disappearing along with the rest, if nothing is done to reverse the process.

Since the 1970s, scientific discourse has been considerably amplified and effectively propagated by the non-governmental organizations (NGO) for natural protection (IUCN, WWF, WRI, etc.), which rallied public opinion around endangered charismatic animals (elephants, whales, pandas, etc.). In the beginning, NGO gave priority to species conservation. Since 1989, they have been collaborating with UNEP

(United Nations Environmental Programme) to develop a global bio-diversity strategy based upon the premise that Nature has an intrinsic right to existence and must be protected from human actions.

1.2.2 'Useful' nature

Ethical and emotional considerations have proven inadequate to rally countries to the cause of biodiversity conservation. Other arguments have been more effective in motivating politicians and policy makers to enact changes. One such approach seeks to demonstrate the utility of biological diversity for the well-being of humankind by citing, for example, the range of cultivatable plants or the therapeutic substances derived from biological diversity. In this context, the term used is 'useful nature', denoting a genetic library that must be preserved to enable the improvement of domestic species.

The Food and Agriculture Organization (FAO) has contributed to this debate with its efforts to promote 'farmers' rights' and recognition of their work in domesticating and improving local varieties of plants and animals. In the fight against famine in the world, the FAO would like to see biodiversity treated as the common heritage of all humankind with free access to resources. But in practice, such initiatives have had little effect. They have been overtaken by the development of biotechnologies and the powerful new roles of industry and national groups, who have their sights on different goals.

1.2.3 Nature has its price

Another consideration, akin to the notion of 'useful nature', is acquiring major significance: the economic interest in biological diversity. On the one hand, naturalists and NGO are enlisting the help of economists to formulate convincing arguments based on the goods and services rendered by biodiversity. On the other hand, the countries involved are beginning to see that industrial interest in biodiversity constitutes a potential source of revenue for patents on forms of life. At the time of the conference in Rio in 1992, the debate polarized around the economic stakes involved in exploiting the value of nature. The first Article of the Convention emphasizes the 'fair and equitable sharing of the benefits arising out of the utilization of genetic resources, including by appropriate access to genetic resources and by appropriate transfer of relevant technologies, taking into account all rights over those resources and to technologies, and by appropriate funding'. Thus, biological diversity is considered as a primary resource for many different kinds of production processes (pharmaceuticals, cosmetics, agricultural foods, etc.). This resource is natural capital that can be exploited and turned to profit.

It will not be long before nations collide over this domain. Most of the resources are in southern hemisphere countries, while the main users, the biotechnology industries, are mostly multinational enterprises of the northern hemisphere. The countries of the southern hemisphere are against the appropriation of their resources without financial compensation and condemn the practice of 'biopiracy'.

1.3   What is Changing?

Since 1993, application of the ratified provisions has altered the situation. By reaffirming the sovereignty of nations over their biological diversity, the Convention confirmed the right of ownership of living things, paving the way for patents and exploitation licenses to be filed, issued and recognised. One might say that at Rio, patent rights emerged victorious over the rights of the environment. This radically transformed the altruistic attitude that had prevailed since the beginning of the 20th century. Biodiversity used to be considered the common heritage of humanity. People were at liberty to exploit the living world and appropriate its derivative forms – the processes and products of its transformation, in accordance with their social position or economic power.

At the same time, there is a heightened awareness that urgent measures must be taken to preserve biological diversity. This is apparent from the number of internet sites and journals addressing the issue, including major scientific publications like *Nature* and *Science*. But at the moment, there is no technical solution to the problem of conservation that meets the needs and is acceptable to society at large. The use and conservation of biological diversity generate fundamental conflicts of interest. Their resolution is contingent upon the choices made by society concerning economic progress and the exploitation of biological resources. For some, priorities may be ethically founded and/or inspired by religious beliefs: we must not destroy that which nature has created over eons of time. For others, the present or potential economic value of biological diversity is sufficient justification to project and implement investments in conservation.

The debate over biodiversity has also given rise to two notions that have grown increasingly popular in recent years: risk and the principle of precaution. We have been confronted with risk in connection with genetically modified organisms (GMO) and the emergence or reappearance of certain diseases which have rekindled old fears: could technological innovations, particularly those involving the manipulation of living organisms, result in a threat to life itself? Some of the objections against the manipulation (and commercialization) of living organisms are voiced in the name of the precautionary principle. The Biosecurity Protocol, signed in January, 2000 in Montreal and ratified in May of the same year in Nairobi, acknowledges the risk that GMO might enter the environment and modify the natural ecological equilibrium. Its goal is to contain risks, even where this is not backed by scientific studies.

The biologists who first raised the issue of biodiversity are no longer the only protagonists of the debate. They are being confronted with a new situation – earlier experienced by atomic physicists in their field – involving continuous, intensive interaction between the progress of scientific knowledge on the one hand and the response of society to emerging perspectives and uncertainties on the other. The question of biodiversity should not remain the domain of one interest group or another; it should rather be regarded as a major problem for society as a whole. There can be no resolution unless all the different protagonists participate. Scientists and socioeconomists must join forces to help clarify the issues.

2 Biological Diversity: What do we Know?

Despite the attention given to biological diversity over the last ten years by both scientists and the media, we are still in no position to draw up an exhaustive inventory – especially as it is not distributed uniformly over the planet. Nevertheless, we do have a sufficiently broad global perception to be able to lay down the foundations for a conservation policy that meets the objectives of the Convention on Biological Diversity.

2.1 The Classification of Living Organisms – Underlying Principles

Classification is a way of organising information by grouping similar taxa. For centuries, we have been trying to describe, name, classify and count species. There are different ways of going about this. Aristotle, in his time, grouped human beings and birds together, because they walked on two legs. Today, classifications are based upon the degree of genetic similarity between individuals, and organisms are grouped according to their phylogenetic relationships.

2.1.1 Levels of organization in the living world

One of the characteristics of the living world is its complex structure and hierarchy: atoms organise themselves into crystals (inanimate world) or molecules, and these molecules, in turn, organise themselves into cells capable of reproduction (living world). Cells can aggregate and co-operate to form multicellular organisms. Individuals – whether single-cell or

Biodiversity   Christian Lévêque and Jean-Claude Mounolou
© 2004 John Wiley & Sons, Ltd   ISBN 0 470 84956 8 (Hbk)   ISBN 0 470 84957 6 (pbk)

Classification

The scientific discipline devoted to naming, describing and classifying living beings is called *taxonomy*. This science is highly formalized and follows the rules of the international codes of nomenclature. *Systematics*, on the other hand, studies the diversity of organisms and strives to understand the relationships between living organisms and fossils, i.e. the degree to which they share a common heritage. What is now called *biosystematics* is a modern approach to systematics that draws upon information from different sources: morphology, genetics, biology, behaviour, ecology, etc.

multicellular – organize themselves into multispecific communities. Taking into account the environment in which organisms live, increasingly complex entities emerge: ecosystems, landscapes and biosphere. On this hierarchic scale, the elements of one level of organization constitute the basic units for the composition of the next, higher level of organization. At each stage, new structures and properties emerge as a result of interactions among the elements of the level below.

- The basic unit of the living world is the *individual*, each bearing its own genetic heritage. The pool of all genes belonging to one individual constitutes its *genotype*. A bacterium contains about 1000 genes; some fungi have around 10 000. Humans have slightly over 30 000.

- A *species* is the group of individuals prone to fertile and fecund genetic exchanges (cf. section 2.1.3)

- A *population* corresponds to a group of individuals of the same biological species inhabiting the same surroundings. It is at this level of organization that natural selection occurs. A species is often distributed over separate populations. Its existence and dynamics are functions of exchanges and replacements among these fragmented, interactive populations, which are called metapopulations.

- Multispecific assemblages that are restricted, usually on a taxonomic basis, constitute settlements or communities. A *biocenosis* is a group of animal and plant populations living in a given place.

- The term *ecosystem* was first introduced by Tansley in 1935 to designate an ecological system combining living organisms with their physical and chemical environment. The Convention on Biological Diversity defines ecosystem as 'a dynamic complex of plant, animal and micro-organism communities and their non-living environment interacting as a functional unit'. This legalistic definition is fundamentally similar to that found in ecological textbooks.

- The biosphere (*sensu stricto*) refers to all living organisms that inhabit the Earth's surface. However, biosphere (*sensu lato*) may also be defined as the superficial layer of the planet that contains living organisms and in which enduring life is possible. This space also comprehends the lithosphere (terrestrial crust), hydrosphere (including oceans and inland waters) and the atmosphere (the gaseous sheath enveloping the Earth).

2.1.2 Taxonomic hierarchies: the search for an evolutionary and functional order in the diversity of species

Classification is concerned with identifying and defining groups or taxa – sets of organisms possessing at least one characteristic in common – and giving them names. A classification of the living world must be hierarchical, because the smaller groups are completely included in larger groups that do not overlap. Initially, in the Renaissance, taxonomy was based on the notion of a descending classification system (the division of large classes into subclasses, as in the classification of inanimate objects). Later, taxonomy shifted to an ascending classification system whereby related taxa are grouped into taxa of a higher order.

Classification of the living world is important for understanding of ecosystems and of biodiversity in general. Postulating that species belonging to the same taxon share a certain number of common biological and ecological characteristics that may differ in those of other taxa, it enables comparisons among species or among taxa of a higher order. Moreover, given that biodiversity is a structural component of the ecosystem, it may sometimes be possible to explain certain ecological functions on the basis of the phyla represented.

In the classification system proposed by Linnaeus, each level of the hierarchy corresponds to the name of a taxon. Naturalists around the world use the same system of general nomenclature – the binominal system – to designate and identify the species. This system consists of a genus name followed by a species name. The superior categories (genus, family, order, division, class, phylum, etc.) indicate the degrees of relationship between taxa (Table 2.1).

The *phenetic hierarchy* is based on the similarity of forms or characteristics among species. According to the premises of *numeric taxonomy*, organisms sharing common characteristics (homologous traits) have similar developmental histories; however, this is not conclusively indicative of their genealogy. Morphological convergences in the course of evolution may have led to possible regroupings. Thus, the Dipneusts (fish with functional lungs, such as *Protoperus*) are morphologically closer to salmon than to cows, but they have a more recent ancestral relationship with cows than with salmon. So how should Dipneusts be classified?

Among today's vertebrates, the group of 'fish' represents a composite class. For example, Actinopterygians (such as trout) are closer to Tetrapods than to Chondrichthians (skates, sharks). As for the coelacanth, this sole known survivor of the group of Crossopterygians is much closer to tetrapods than to other groups of fish, with the exception of Dipneusts, another very ancient group currently represented by several species, such as for example, the African *Protopterus*.

The *phylogenetic hierarchy* is based on the evolutionary relationships of groups descending from common ancestors. The cladistic classification

Table 2.1 Hierarchic biological classification of three animal species

Level	Species 1	Species 2	Species 3
Domain	Eukaryotes	Eukaryotes	Eukaryotes
Kingdom	Animal	Animal	Animal
Phylum	Arthropods	Arthropods	Chordates
Class	Insects	Crustaceans	Mammals
Order	Diptera	Decapods	Primates
Family	Nematocera	Caridae	Hominidae
Genus	*Aedes*	*Homarus*	*Homo*
Species	*aegypti*	*americanus*	*sapiens*

system (sometimes called the Hennigian system) is based on the principle that during the course of evolution, an ancestral species gives birth to two daughter species. If one takes three species and compares them two-by-two, the pair that has the more recent common ancestor is grouped together. A group of species is termed *monophyletic* if it derives from a single common ancestor, while a *polyphyletic* group comprises species that manifest similarities but are not all directly descended from a common ancestor.

The methods of molecular phylogeny are also founded upon the hypothesis that resemblances between two organisms will be more numerous if their ancestral relationship is closer. Here, however, genetic sequences are compared rather than morphological traits. Thanks to the methodological advances of molecular biology, phylogenetic classification is currently progressing beyond phenetic classification.

As research advances, it becomes possible to modify classifications to reflect different lines of development with greater accuracy. Alterations are frequent for both currently existing species and fossil species. The science of taxonomy is in a state of perpetual evolution. The Eukaryotes, for example, constitute a highly heterogeneous group within which four major categories are currently distinguished: animals, plants, fungi and a poorly defined group called protists. Because their molecular phylogenies have revealed hitherto unimagined relationships, the relationships between these Eukaryote groups are currently undergoing intense revision. For example, for a long time, microsporides were considered primitive because they lack mitochondria, while in fact they represent later lineages that happen to have lost their mitochondria. Even stranger: several independent studies attest to a relationship between fungi and metazoa that excludes plant species. Therefore (all things considered equal...), we are more closely related to truffles than to daisies!

The classic view of metazoa in the form of a phylogenetic tree with increasing complexity (Figure 2.1a) has been replaced by a view in which all the 'intermediary' groups have disappeared (Figure 2.1b). Arthropods would thus appear to be closely related to Nematodes, while molluscs belong to a group comprising Brachiopods, Annelids and Platyhelminthes. As for insects, they are probably crustaceans that have adapted to terrestrial habitats. Molecular phylogeny reveals that the apparent simplicity of certain groups long considered to be primitive (Platyhelminthes, Nematodes) is probably due to secondary simplification.

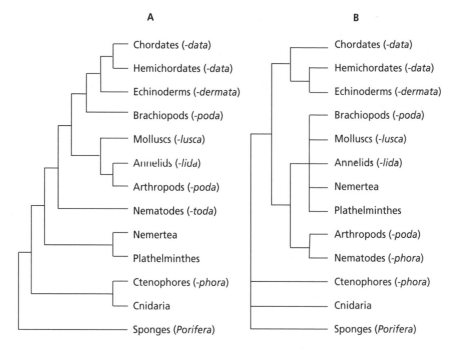

Figure 2.1   Comparison of two phylogenies of metazoa:
A. Traditional phylogeny, where successive clades appear in the order of increasing complexity
B. Phylogeny obtained by using sequences of 18S rRNA and Hox genes

2.1.3   The concept of species

The concept of species has long been an object of controversy, and to this day, there is no completely satisfactory definition. Until the mid-18th century, systematicians had a fixed conception of species: they were such as God had created them and limited in number. Accordingly, the aim of taxonomy was to compile an inventory of all the existing forms of life and describe their specific characteristics. Linnaeus formalized this concept by defining each species in terms of a single type (holotype). A species was constituted by the sum of individuals identical to each other and to its 'type' specimen. In other words, the sample specimen served to describe and characterise the species in morphological terms. Sample specimens were stored in a museum for future reference or as a sort of standard for later comparisons.

This fixed notion could not withstand the discovery of the mechanics of evolution (mutation, selection, genetic drift), and towards the mid-20th century, it gave way to the concept of dynamic biological species, founded not only upon resemblance but also upon the interfecundity among individuals constituting a population and among their descendants. Whilst the donkey and the horse can reproduce, they remain distinct species because their descendants are infertile. It is the reproductive isolation of a group of individuals that defines them as a species; however, demonstrating interfecundity is another matter. Because it is physically impossible to cross the majority of wild organisms so as to establish or refute their potential interfecundity, the concept of biological species is obviously difficult to apply. Besides, this definition can only be strictly applied to species that engage in bisexual reproduction, leaving the question of micro-organisms up in the air. Thus, despite certain reservations, species continue to be identified primarily by morphological descriptions wherever possible, complemented by a biochemical description, such as in the case of bacteria.

Within one species, it is possible to recognize various subunits that are considered as subspecies, races, strains, varieties, etc. There are no precise and universally accepted definitions for these intraspecific categories, which may be based on morphology, geography or genetics. Among the numerous races of domestic animals, we observe forms that are highly differentiated in morphological terms. However, intraspecific variability can also be expressed in other ways, for example in reproductive behaviour or modes of communication (see box on the common finch).

The new tools of molecular biology are a valuable aid in distinguishing individuals belonging to species that are morphologically very close: this is the case for *sister species*, which are true biological species that have achieved reproductive isolation but are still difficult to distinguish solely on the basis of their morphological characteristics.

It is also now possible to investigate intraspecific genetic variability and the relationships among individuals with a greater degree of precision. Individuals within a population actually have slightly different genotypes. This genetic polymorphism can be quantified in terms of allelic frequencies that vary from one population to another and develop over time. In phylogenetic terms, species may be defined as a single lineage of ancestor-descendent populations, which is distinct from other such lineages within its range and evolves separately from all lineages outside its range.

The intraspecific variability of the common finch

In southwest France, the common finch (*Fringilla coelebs*) uses around 50 distinct dialects that are regional variations of the same language. All the dialects used by this species are 'authentic,' in contrast to certain birds capable of learning new songs throughout the course of their lives, thus producing 'false' dialects. In its first year, the fledgling bird receives warbling lessons by listening to its parents or close neighbours; afterwards, its repertory will never change, no matter what the circumstances. Thus, analysis of existing finch dialects yields a relatively precise map of the subpopulations inhabiting the same region.

The results of a study conducted in the southwest provide evidence of three major groups: located in the Massif Central, the Pyrénées and the Landes regions, respectively. These super-dialects are fundamentally different, sharing no songs in common. A closer study of their distribution reveals that groups that are close geographically sing some very different songs, while others, that are geographically farther apart, sing more closely related songs. Fragmentation into subpopulations is not a regular phenomenon over space.

Every species that is valid in 'biological' terms will be recognized within this framework. But different populations of the same species will also be identified as distinct entities on the basis of their genetic properties, even though they are not isolated in reproductive terms. What criteria should be used to separate and define subgroups? We are again faced with the problem of classifying forms or varieties, similar to that which arose for morphological taxonomy.

2.1.4   Ecosystems

The concept of ecosystem is a rather abstract notion: a chemical and physical environment (biotope) is associated with a community of living organisms (biocenosis), and together they set the stage for a system of interactions among the constitutive elements. In practice, however, ecologists tend to define ecosystems as geographically defined entities such as lakes, watersheds or mountain ranges.

Ecosystem functioning is characterized by:

- *flows of energy* between organisms such as plants that accumulate solar energy through photosynthesis, herbivorous animals that utilize this energy, and decomposers that recycle organic matter;

- *biogeochemical cycles* circulating matter in the form of mineral or organic substances. Such cycles apply in particular to water, carbon, oxygen, nitrogen, phosphorus, etc.;

- *food chains* that impose a trophic structure upon the ecosystem. Trophic – or alimentary – interactions are the driving forces for the flows of energy and matter.

The concept of ecosystem is intrinsically dynamic: flows, biogeochemical cycles and trophic structures are continuously evolving over time and

---

Biocenoses: structured or fortuitous assemblages?

A core question of ecosystem ecology is whether the whole assemblage of species observed in a particular place is a fortuitous collection of populations that have succeeded in colonizing the ecosystem and maintaining themselves, or else a selection of co-evolved species that established a network of interdependencies over time. Many ecologists currently tend towards the latter theory, but they are having a good deal of difficulty substantiating these different types of interaction.

Actually, the time factor plays an important role. When a new habitat is created, it is colonised by opportunistic species, and its settlement is largely fortuitous. With time, there may be a co-evolution of species, and a greater degree of interdependence may develop. For example, Biocenoses: structured or fortuitous assemblages? the temperate northern lakes, which were frozen over during the last ice age 15 000 years ago, were recolonized from nearby surroundings after the ice melted. As a result, their fauna is not very diversified and there are no endemics present. Conversely, the great lakes of east Africa, that have existed for millions of years, have a fauna which is rich in species and endemics, with complex interspecific relationships resulting from a long period of co-evolution.

space. A good example to illustrate this phenomenon is the river, with its lowest water level and alluvial plain: in the course of the hydrological cycle, the spatiotemporal dynamics of flooding profoundly modify the landscape as well as the interactions among species.

The biosphere is the ultimate ecosystem. Recognition of the importance of global factors (natural and human-induced climate changes, major biogeochemical cycles, globalization of the transfer of species, etc.) has stimulated scientific interest in this level of organization. Research on the global functioning of the ecosystem Earth has become a reality.

2.2.   The Inventory of Species

The biological diversity of living organisms is evident at all levels of organization, from genes to ecosystems. But the term usually refers to the diversity of species, based on a tally of those occupying a given space for a given amount of time.

Around 300 years ago, botanists and zoologists set out to describe and inventory living organisms. In the mid-18th century, Carl Linnaeus enumerated 9000 species of plants and animals. Two hundred and fifty years later, with over 1.7 million species described, we know that the inventory of the living world is far from complete, especially in tropical regions. Nobody really knows how many different species live on Earth, but their number is estimated at somewhere between 7 and 100 million (Table 2.2). The uncertainty is indicative of the extent of our ignorance, which is particularly bothersome when trying to demonstrate that human activities are causing an unprecedented erosion of biological diversity. At an average rate of about 10 000 new species described each year, it will take several more centuries to complete the inventory.

Our knowledge varies with taxonomic groups. More or less exhaustive censuses are only available for a small number of botanical or zoological groups, such as mammals and birds, of which over 95 per cent are currently known. By contrast, the number of insects is certainly higher than the already considerable number (950 000) registered so far. Insects account for close to two-thirds of the new descriptions of recent years. As for fungi, their total number is probably between 1 and 2 million, while nematodes, little parasitic worms living on plants and animals, run to several hundreds of thousands. The sources for newly discovered species are mainly situated in the tropics, in coral reefs and deep sea beds, but

Table 2.2 Approximate number of species currently recorded and estimated number of existing species for vertebrates and other groups of plants and animals presumed to include 100 000 or more species

The number of probable species is a hypothetical extrapolation, but it gives an idea of the order of magnitude of the richness of the living world.

Taxonomic groups	Approximate number of recorded species	Estimated number of species
Viruses	4000	500 000?
Bacteria	4000	1 000 000?
Fungi	72 000	1 to 2 million?
Protozoa	40 000	200 000?
Algae	40 000	400 000?
Plants	270 000	320 000?
Invertebrate animals	100 000	10 000 000
Sponges	10 000	
Cnidaria	10 000	
Platyhelminthes	20 000	
Nematodes	25 000	400 000
Arachnids	75 000	750 000
Crustaceans	40 000	150 000
Insects	950 000	8 000 000
Molluscs	70 000	200 000
Annelids	12 000	
Echninoderms	6000	
Vertebrate animals		
Fish	22 000	25 000
Amphibians	42 000	4500
Reptiles	6500	6500
Birds	9672	*Idem*
Mammals	4327	*Idem*

also found in all latitudes in regions that are difficult to access, as well as in the realm of small species (soil fauna, marine meiofauna) and parasites. One cubic metre of temperate prairie soil contains thousands of species of micro-organisms and invertebrates whose taxonomic attribution and metabolic activity is largely unknown.

For other groups such as bacteria and viruses, where species are more difficult to characterize than in the case of vertebrates or insects, real numbers certainly far exceed current estimates (cf. box on molecular ecology). Using molecular biology techniques, it has recently been shown that marine picoplankton (tiny organisms measuring 0.2–3 μm, the basis for food webs in pelagic ecosystems) is composed of numerous groups of unregistered eukaryotes. In the Pacific Ocean, for example, most of the ribosomal RNA sequences obtained from picoplankton samples belong to unknown organisms. It is interesting to note that most of these sequences belong to predatory species or species involved in the decomposition of organic matter – a function which was hitherto primarily associated with bacteria and not with eukaryotes. Moreover, primitive species of green algae (Prasinophytes) have been discovered that had never previously been isolated.

Molecular ecology

For a long time, the notion of species applied to bacteria was also based on the number of physiological and biochemical characteristics held in common by a type of strain. But molecular biology enables the comparison of gene sequences, paving the way for phylogenetic approaches to such questions. The most frequently used molecule is the tiny subunit (16S) of ribosomal RNA. In addition, the PCR (polymerase chain reaction) method of amplification has made it possible to work with the bacterial DNA of wild species without resorting to cultures. PCR enables speedy determination of genetic sequences on the basis of a small number of cells – or even a single cell. An entire inventory of molecular diversity can be generated in this way, revealing the enormous diversity of prokaryotes in all environments, including those with extreme physical and chemical characteristics. These studies furthermore show that a large proportion of this diversity does not relate to any known lineage; rather, a number of significant, previously unknown groups exist that are genetically well differentiated from one another. In many environments, it is only possible to isolate and cultivate a tiny proportion of the species actually present.

2.3   Systematics, Computer Technology and the Internet

Naming, classifying and identifying species is a delicate task requiring the use of:

* collections of type specimens, which are, in principle, deposited in museums;

- specialized publications describing new species;

- flora and fauna, accompanied by identification keys that synthesize available information and render taxonomic knowledge accessible.

As a result, knowledge of the diverse taxonomic groups long remained the privilege of a handful of specialists whose numbers tended to fluctuate with changing politics and fashions. Today, we may be facing a true crisis in the recruitment of taxonomists. We must accelerate the process of inventorying biological diversity, which will require several more centuries, whilst making better use of existing information, which is often dispersed over any number of conservation programmes. Given the situation, powerful and interactive means are essential to the management and dissemination of taxonomic information.

Computer technology had naturally become an indispensable tool for storing, managing and analysing all this information. Despite considerable advances in personal computers and computer processing, there is still no effective computer information system to deal adequately with questions of systematics and taxonomic inventories.

Some developments are underway: taxonomic indexes, consisting of lists of reference names and their synonyms, are being constructed for access on the internet. They establish the use of names, often supplemented by information describing the species, its geographic distribution, collections stored in major museums, etc. For example, the fish collection of the National Museum of Natural History in Paris is accessible on the internet. Among the various international efforts to promote the dissemination of knowledge in the field of systematics, one might mention the international Species 2000 project, the European Fauna project, the European Register of Marine Species (ERMS), or the Check List of European Marine Mollusca (CLEMAM).

Some of these databases also include systems to aid professionals and amateurs in the identification of species. We are currently witnessing the development of a computer-based taxonomic support programme, enabling the user automatically to construct identification keys, undertake phylogenetic diagnoses or reconstructions, establish identities, and store and access data from various sources, including collections kept in museums.

Generally speaking, the objective is not to construct huge, isolated databases, but rather to use the internet to organize how knowledge is shared (shared databases and computer facilities, cooperative work, etc.) and to ensure compatibility and synergy amongst the different initiatives that are underway. This is the objective of the GBIF (Global Biodiversity Information Facility) in the Megascience Forum of the OECD, which hopes to develop into a major source of information on biological diversity – a kind of super-catalogue of life. In terms of providing and using the huge quantity of systematics information that already exists, current initiatives may still be far from adequate; nevertheless, rapid progress is being made.

2.4   Measuring Biological Diversity

Quantifying biodiversity is of practical importance when considering its evolution over time, geographical zones of interest and conservation strategies.

Different methodologies are adopted for measuring biodiversity. None of them is universally accepted, and the choice of methods and scales tends to depend upon the objective pursued. From a theoretical standpoint, the correct procedure would be to evaluate all aspects of biodiversity in a given system. But such a task is practically impossible to accomplish. We must make do with estimates based on a number of indicators, including genetics, species or populations, the structures of habitats, or any combination that provides a relevant, albeit relative, evaluation of biological diversity.

The most common unit of measurement is the species richness determined for all the taxa, or subsets of taxa, identified in a given environment. However, confusion should be avoided between biodiversity and species richness: the former includes the latter but is not restricted to it. A high number of species in a given environment is likely to be a good indicator for a larger genetic, phylogenetic, morphological, biological and ecological diversity. In groups with well-known taxonomies, the list of species is relatively easy to establish. For others, it is more difficult.

The relative density of each species (also known as 'evenness') has also been used to compare different communities or ecosystems. The most frequently used indices are based on the estimated relative abundance of the species found in the samplings. However, these indices assign an equal functional weight to all species, for which there is no clear justification. Other indices have therefore been developed, taking into consideration such factors as taxonomic position, trophic state, or body size of the species. Generally speaking, the usefulness of such indices is limited, because they do not provide much information that is relevant on a practical level. Attempts have been made to enhance them with genetic and ecological input. Analogous indices are used in genetics: e.g. richness (the number of alleles for the same locus), evenness (the relative frequency of alleles), and heterozygosity, which associates the number of alleles with their relative frequency. Another approach involves identifying the diversity of ecosystems in a landscape, or habitats within an ecosystem. It is possible to proceed as in taxonomy by identifying, naming and classifying entities, comparing different situations, and then attempting to generalize one's observations. This typological approach has established several categories of classification based on floral or faunal characteristics, on assemblages of species (phytosociology), or on landscape features (ecoregions, phenological structures, etc.). One example for a typology of habitats is the classification system of European habitats CORINE.

2.5   The Geographic Distribution of Biological Diversity

Biological diversity is not evenly distributed over the surface of the planet. Naturalists have attempted to determine large-scale tendencies or patterns in the spatial distribution of biological diversity. Plotting characteristics of climate against vegetation has long been used as a means of identifying large biomes (Figure 2.2). Alternatively, assessing the degrees of relationships between flora and fauna led to the delineation of biogeographic areas. In each case, the typological process subscribes to a hierarchical system, subdivided according to the degree of accuracy required. Alternatively, scientists can try to identify distinct areas that are particularly rich in endemic species.

Endemic species

A species is considered endemic to an area if occurs there and nowhere else. Usually, endemicity is the consequence of the geographic isolation of taxa that evolved independently. The area of endemism can either be relatively large (three-quarters of the Malagasy mammalian species are endemic), or else restricted to a well identified ecosystem, such as the hundreds of cichlid fish species that inhabit the great lakes of east Africa (Lake Victoria, Lake Malawi, Lake Tanganyika).

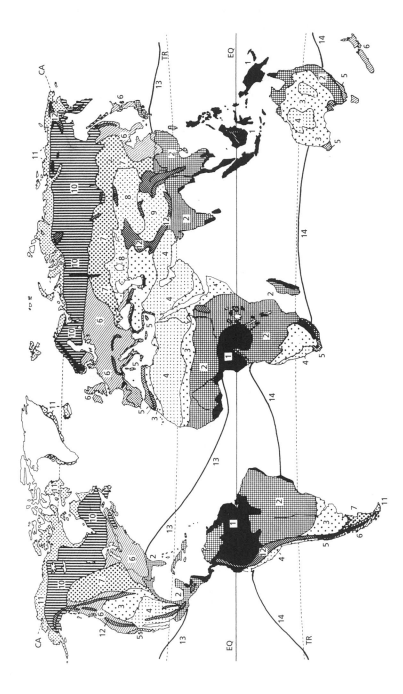

Figure 2.2   Distribution of the principal biomes (based on vegetation patterns) over the surface of the Earth (according to Ozenda, 2000) 1. Equatorial rainforest 2. Tropical rain forest 3. Dry savannas and steppes. 4. Desert 5. Mediterranean sclerophyllous forest 6. Temperate deciduous forest 7. Cold continental steppes 8. Cold Asiatic deserts 9. High-altitude deserts 10. Boreal coniferous forest 11. Tundra 12. Mountain ecosystem 13. Northern limit of coral reefs 14. Southern limit of coral reefs

2.5.1 The taxonomic diversity of marine environments is greater than that of terrestrial environments

Many more animal phyla exist in the marine world than in the terrestrial world (Table 2.3). Just one, the Onychophora, is restricted to terrestrial environments, whilst two-thirds of all the rest are restricted to marine environments.

Table 2.3 Distribution of major metazoan groups, comparing richness in species by type of habitat (According to May, 1994)

Group	Benthic marine	Pelagic marine	Freshwater	Terrestrial
Annelids	***	*	**	***
Arthropods	***	***	***	***
Brachiopods	**			
Bryozoans	***		*	
Chaethognaths	*	*		
Chordates	***	***	***	***
Cnidaria	***	**	*	
Ctenophores		*		
Echinoderms	***	*		
Echiuroids	**			
Gastrotrichs	**		**	
Hemichordates	*			
Kamptozoans	*		*	
Kinorhynques	**			
Loricifers	*			
Molluscs	***	**	***	***
Nematodes	***		***	***
Nemerteans	**	*	*	*
Onychophorans				*
Phoronidiens	*			
Placozoans	*			
Platyhelminthes	***	*	***	**
Pogonophorans	**			
Sponges	***		*	
Priapulids	*			
Rotifers	*	*	**	*
Sipunculiens	**			*
Tardigrades	*		**	*
Total	26	11	14	11
Endemic	10	1	0	1

* <100; **<100–<1000; *** > 1000

However, just under 15 per cent of known species are marine, despite the proportionately greater size of the ocean surface. There are two possible explanations for this that are not contradictory: first, marine environments have been less extensively explored than terrestrial environments; and second, marine environments are more homogeneous and provide less opportunity for speciation. Phyla that have colonized the terrestrial world diversified more, the most obvious example being insects, which are phylogenetically 'derived' from crustaceans.

2.5.2   Gradients of spatial distribution

In their attempt to ascertain a natural order, scientists have tried to pinpoint factors that might explain the spatial distribution of species currently observed. The identification of gradients is one way to achieve a clearer understanding of the organisation of biological diversity.

2.5.2.1   Latitudinal gradients

A general phenomenon of terrestrial and continental aquatic environments is the existence of a latitudinal gradient: for most taxonomic groups, species richness increases from the poles towards the Equator. In other words, biological diversity is much greater in the tropics than in temperate regions (Figure 2.3a). This phenomenon is especially conspicuous in the case of plants. However, for other groups, such as soil nematodes, there seems to be an inverse situation, with greater richness at high latitudes.

The existence of a latitudinal gradient in marine environments has been established for pelagic ecosystems, as well as for hard substrates' benthic fauna. But the phenomenon is debatable for other groups, and sometimes, the reverse even appears to be the case. Thus, macro-algae are more diversified in temperate than in tropical regions. The same is true of marine birds that feed on fish and crustaceans. This could imply that various groups of marine organisms are not sensitive to the same ecological factors as organisms inhabiting the terrestrial environment. No clearly delineated latitudinal gradient is in evidence in the southern hemisphere, and the richness of marine species in Antarctica is particularly high.

Of course, attempts have been made to explain the latitudinal gradients of terrestrial environments by the fact that tropical regions occupy a larger surface area than temperate or cold regions. The existence of a

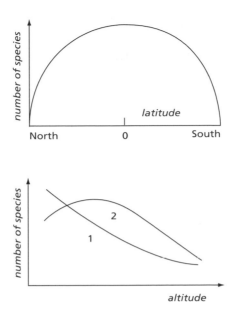

Figure 2.3 (a) Schematic curve of the distribution of species richness at different latitudes. This is the pattern observed in particular for amphibians, reptiles, etc. (b) Changes in species richness as a function of altitude 1: Graph of the decrease in species richness observed for bats in the national park of Manu (Peru); 2: Graph of the bell-shaped distribution observed for the terrestrial birds of South America

much greater proportion of endemic species in the tropical than in the temperate zone is probably also a consequence of climatic variability. Successive glaciations and deglaciations, occurring approximately every 100 000 years during the Quaternary, were extremely disturbing to the cold and temperate zones, where they had an 'ice-scraper effect' upon biological diversity. By contrast, tropical climate zones were conducive to long-term ecosystem perenniality, despite occasional large-scale perturbations. Seasonal alternation also seems to play a key role: because climatic variability is much greater in temperate zones, species are compelled to undergo costly adaptations (to frost, for example). The relatively stable conditions found in tropical zones, both in seasonal terms and over longer periods of time, allowed numerous organisms to specialise and occupy the different ecological niches available in the regions.

Further hypotheses have been advanced: for example with greater solar energy input, tropical regions are more productive, facilitating the coexistence of numerous species.

2.5.2.2   Longitudinal gradients

In the marine domain, there is a well-established longitudinal gradient for the diversity of corals, with the greatest species richness observed in the Indonesian archipelago. It diminishes westward, with certain irregularities in the Indian Ocean (and with the exception of the Red Sea, for certain groups), and drops to its lowest level in the Caribbean.

2.5.2.3   Altitudinal gradients

Altitudinal zoning is a form of organization for biological diversity encountered in mountainous regions, where it is possible to observe rapid changes in settlement structure over a limited surface area as the altitude increases. Temperature and rainfall are the main physical factors defining this gradient. For some taxa, species richness simply diminishes with rising altitude, while for others, species richness takes the shape of a bell curve (Figure 2.3b).

2.5.2.4   Depth

In the sea, scientists distinguish between the *pelagic* domain, covering species and communities that live in open waters, and the *benthic* domain, for organisms living on and in the sediments or on hard substrates. Generally speaking, biological diversity is greater in benthic surroundings than in pelagic surroundings, and greater in coastal regions (where there is a greater diversity of habitats) than in the high seas. It is organized around several major domains.

- The *continental plateau* is the coastal zone extending out to an average depth of 200 m. Called the *neritic* zone, it is home to the bulk of benthic organisms. In tropical regions, this zone also comprises coral reefs, which, to draw an analogy, are to marine biodiversity what tropical forest systems are on land.

- The *continental slope* runs along the rift between edge of the continental plateau and the oceanic domain. This is the *bathyal domain*, where it is possible to investigate gradients of species richness over depth. As a general rule, maximum species richness lies within the range of 1000–1500 m for pelagic communities and 1000–2000 m for megabenthos. Such gradients may result from the combined effects of depth and latitude.

- The ocean basin is formed by the abyssal plain at 4000–6000 m below sea level. It has deeper troughs as well as mid-ocean crests (2000–3000 m). Until quite recently, life in the ocean depths was held to be scarce, but evidence has emerged that deep-sea sediments are home to a large variety of species, the majority of which have not yet been described with any degree of precision. Scientists have also observed various fauna and most singular forms of microbial life that derive their energy from nearby hydrothermal sources through a process of chemosynthesis. Oceanic ridges have furthermore revealed ample fauna with many endemic species, especially in the Pacific.

2.5.3 The relationship between surface area and species richness

There is an empirical relationship that is well known to ecologists between the surface of an island and the number of species observed on that island. This relationship is usually expressed by the Arrhenius equation: $S = cA^z$, where S is the number of species, A is the surface, and c and z are constants. The island biogeographic theory applies equally well to ocean islands and to continental islands, such as mountain summits, lakes or watersheds that are isolated from one another.

The relationship between the surface area of an insular system and the species richness of a group has been demonstrated on many occasions, although there are deviations resulting from the different histories of biogeographic regions. The example presented in Figure 2.4, which looks at African and European rivers, shows that for every region, there is a clear relationship between the surface of a river catchment (or the discharge of a river) and its species richness in fish.

The diversity of habitats is also likely to increase with an increase in surface area, and settlements to grow richer in species as habitats become more diverse.

A further assumption is that the number of species observed on an island is the result of a dynamic equilibrium between the natural extinction of species and the rate of immigration of species from a continental source that is richer in species. This is the theory of dynamic equilibriums advanced by MacArthur and Wilson (1963). On a continental scale such as Europe or Africa, extinctions are compensated for by processes of speciation.

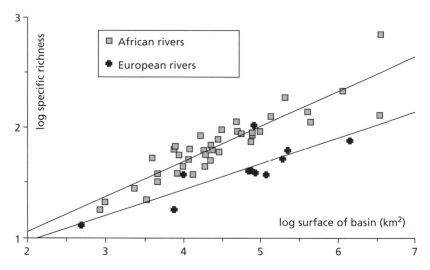

Figure 2.4 Species richness and surface of river catchment: comparison between European rivers and African rivers

2.5.4   Biomes as ecological units

The distribution of species over the surface of the Earth is not random. It results from a combination of ecological factors, interacting with the preferences and abilities of organisms. On the basis of the combined factors of precipitation and temperature, the Earth can be divided into large morphoclimatic domains. On an extremely macroscopic scale, four ecoclimatic zones can be identified: tropical, hot and humid; temperate humid; polar; and arid. On a more differentiated scale, it appears that different regions with identical climatic conditions are occupied by comparable natural ecosystems. Vegetation has the virtue of being a quite reliable indicator for plotting the interplay of such diverse factors as geomorphogenesis and climate on rather large spatial scales. The boundaries of large vegetation formations mark the discontinuities apparent in the natural world. Homogeneous in climate (temperature and precipitations), biomes are macrosystems on a regional scale (Figure 2.2).

The isotherm of 10°C for the warmest month of the year, roughly corresponding to the septentrional limit of the great forest of conifers in the northern hemisphere, separates the cold regions of the high latitudes from the temperate regions of the medium latitudes. Similarly, the isotherm of 20°C for the coldest month of the year quite accurately delineates those regions characterised by a constant temperature. Within this vast domain, moisture patterns differentiate the great pluvial forest, which is continuously hot and humid throughout the year, from the savannas, with their characteristic dry season.

How many biomes are identified depends upon the desired degree of resolution, and different authors distinguish anything from 10 to 100. In most cases, the physiognomy of the vegetation serves as the basis for delineating biomes: forests (24 percent of the land surface), savannas (15 percent), prairies and tundras (15 percent), etc. One should also mention cultivated areas, which cover over 10 percent of the land above sea level, as well as the deserts and frozen expanses (30 percent).

2.5.5 Biogeographic regions as taxonomic units

Numerous attempts have been made to divide the surface of the Earth into large biogeographic regions by proceeding from the current distribution of flora and fauna and knowledge about their historical locations and drawing inferences as to the spatial organisation of biological diversity. Very broadly, six major terrestrial regions are recognized within which the flora and fauna share a common history. Corresponding globally to the major continental shelves, these include three large 'tropical' regions: the Afrotropics (Africa), the Neotropics (South America) and Indo-Malaysia, which are home to over two-thirds of the known terrestrial species. The Nearctic region (North America), the Palearctic region (Eurasia; see Table 2.4) and the Australian region (Australia) cover the temperate to cold zones.

This typological approach can be further refined. Within each of the six major zones, any number of subsets may be identified depending upon the

Table 2.4 The species present in Europe as compared with the world

	Known species in Europe	Species known only in Europe	Number of species known in the world
Freshwater fish	344	200	10 000
Reptiles	198	90	6500
Amphibians	75	56	4000
Nesting birds	520	30	9700
Mammals	270	78	4327
Vascular plants	125 000	3500	270 000

accuracy required and criteria applied. Thus, continental Europe has been divided into several biogeographic regions or domains (Figure 2.5).

France is the only country in Europe with large territories in four very different biogeographical domains: Atlantic, continental, alpine, Mediterranean. This explains the great diversity of vertebrates noted in this hexagon. The same holds true for natural habitats: from the pine forests of the Jura Mountains (an almost boreal climate) to the steppes of the Crau near Marseille, home to species with a strong African affinity. Thanks to its ecological diversity, France has been able to weather the destruction of flora and fauna connected with the Industrial Revolution as well as the transformations incurred by agriculture: deforestation, drainage of wetlands, the cutting of hedgerows, cultivation, the battle against 'pests,' etc. Only a small number of species disappeared before the 20th century, but the number of endangered species has grown with the magnitude of human impact upon the environment.

2.5.6   'Hotspots': areas of high diversity

Great importance has been attributed to identifying geographic zones on the basis of their overall specific richness or richness in particular endemic species. Some researchers have identified what they call hotspots of biological diversity: zones that are blessed with an extraordinary concentration of species but are at the same time subject to an accelerated loss of habitats. Distributed around the Earth and obviously of critical significance to conservation, these hotspots are threatened by human impacts. Areas that are rich in endemic species are areas where a large number of taxonomic groups have been able to survive over the long term. Some believe that these are 'Pleistocene refuge zones,' where biological diversity was able to maintain itself during the ice ages.

One study has shown that 44 percent of all vascular plants (i.e. more than 130 000 plants) and 35 percent of all vertebrates except fish (i.e. around 10 000 species) are confined to 25 hotspots, covering only 1.4% of the land surface of the Earth. Most of these sites are in tropical zones; however, five are in the Mediterranean basin (Figure 2.6), and nine are islands – among them, Madagascar, which is home to over 11 000 higher plants, with an 80 percent rate of endemicity. By contrast, the number of endemic species in Europe represents a mere 2–6 percent of global species in morphological taxonomic terms.

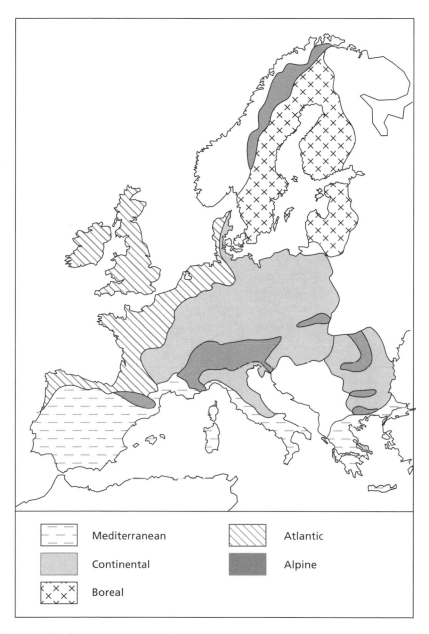

Figure 2.5 Simplified illustrative map of the biogeographic regions of Europe This map was drawn by the European Topic Centre on Nature Protection and Biodiversity for the European Commission initiative, 'Natura 2000'. It covers only the 15 member states of the European Union and the 12 candidates for entry

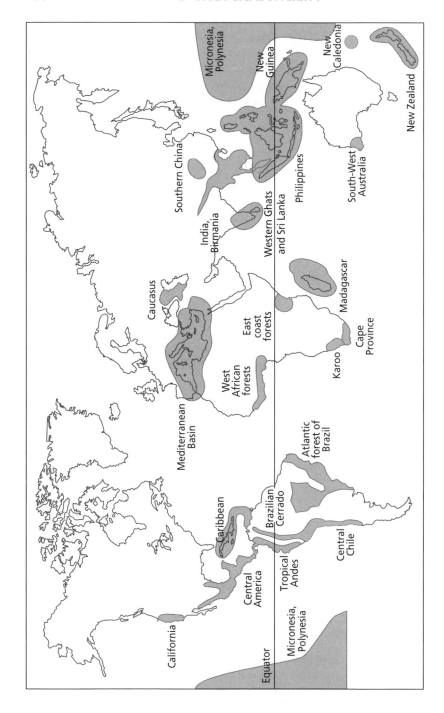

Figure 2.6 The 25 hotspots manifesting an exceptional biological diversity and at the same time endangered by human activities

3 The Mechanisms at Work in the Diversification of Life

The diversity of the living world is a source of unending wonder. Hypotheses explaining its origins have developed considerably over the past two centuries. Toward the end of the 18th century, Cuvier was an advocate of creationism (living beings are faithful replicas of God's creations), while Buffon defended the theory of spontaneous generation. The latter view was, by and large, shared by many other civilisations, whether in China, India or in Egypt. It was Lamarck, at the beginning of the 19th century, who developed the idea of species transforming into other species. But Lamarck's *transformism* is tinged with finalism: there exists an 'inner force' which propels organisms to adapt to changes in their surroundings, and these transformations are transmitted to their descendants. In *The Origin of Species*, published in 1859, Darwin borrows certain ideas from Lamarck but rejects his finalism in favour of the principle of *natural selection*: those individuals who are best adapted are selected through competition. Notwithstanding his advanced insights into evolution, Darwin did not, at the time, provide irrefutable proof to support his ideas on the origin of life.

Experimenting with sterilisation toward the end of the 19th century, Pasteur demonstrated that life cannot be generated spontaneously from inanimate matter. Around 1900, the rediscovery of Mendel's laws inspired the study of genetics and accorded an active role to mutations in the process of evolution. Around the mid-20th century, the *synthetic theory of evolution* integrated natural selection and mutation into one process. The unit of evolution is the population, and natural selection benefits populations that adapt to their life circumstances by favouring the transmission to their descendents of alleles that are advantageous to the species.

Biodiversity   Christian Lévêque and Jean-Claude Mounolou
© 2004 John Wiley & Sons, Ltd   ISBN 0 470 84956 8 (Hbk)   ISBN 0 470 84957 6 (pbk)

3.1   What is Life?

'What is life?' Humankind has been asking this question for a long time. Religions and philosophies have proposed their own definitions, but scientists have always found it difficult to deal with such questions. What conditions are necessary and sufficient for life to appear and for a system to qualify as living? In fact, the borderline between inorganic matter and the most primitive forms of life is not always clear. Current efforts to revise explanatory hypotheses are turning to modern chemistry for new insights.

Some consider life to exist as soon as a molecule is capable of replicating itself. According to the exobiology programme of NASA, life is a self-maintaining chemical system capable of Darwinian evolution. For others, the most elementary organisms must be at least enclosed by a membrane to qualify as life. The simplest living objects known on Earth today are cells, and life, as we know it, is essentially cellular. We know of no natural life forms corresponding to isolated molecules capable of auto-replication. However, we cannot exclude the possibility that such forms may have existed a long time ago, at the origins of life.

3.1.1   The chemistry of the origins of life

Life results from complex chemical processes basically involving the interaction of three types of macromolecules capable of transmitting information: proteins, ribonucleic acid (RNA), and deoxyribonucleic acid (DNA). These are long linear polymers whose monomers are amino acids in the case of proteins, and nucleotides (bases) for nucleic acids. The genetic message corresponds to the sequence in which the four bases (adenine, A; cytosine, C; thymine, T; guanine, G) are arranged along the DNA or RNA molecules.

3.1.2   How did life appear on Earth?

Despite the progress of science, the circumstances of life's appearance upon Earth are still unclear. We remain ignorant as to the possible origins of prebiotic molecules, whose chemical evolution may have resulted in the polymers characteristic of life on Earth. Palaeontologists have no fossils dating from the origins of life, but chemical theorists describe possible scenarios. The universality of the genetic code suggests a common origin for all known living beings of cellular nature. A plausible hypothesis is that life began as the result of multiple chemical reactions in inorganic matter that chanced to set off a cascade of events, permitting certain molecules to reproduce and to organise themselves into living beings.

The primitive atmosphere of Earth contained large quantities of carbon dioxide (CO_2), molecular nitrogen (N_2) and water vapour. Although the sun was still pale (the solar energy reaching Earth was only 75 per cent of the current level), the climate was very warm due to the greenhouse effect. With temperatures around 100 °C, the ocean waters were highly enriched with a variety of molecules constituting a 'primitive soup'. Certain atoms contained in this soup, such as C, H, O, N, P, S, would have been at origin of all organic molecules on Earth today. Under the impact of various sources of energy, such as heat, lightning and ultraviolet solar rays, mineral matter formed the first organic molecules: some protein-based amino acids, some fatty acids based on lipids, nucleotides, etc. It is likely that the presence of numerous metallic minerals dissolved in the oceans (Fe, Mg, Mn, Ca, Cu, Mb, Zn, Co, etc.) facilitated the catalysis of the reactions by which organic molecules were formed.

There is a consensus that the scenario of the origin of life took place in water or at the interface of an aqueous phase and an organized chemical substrate. Deposits of organic molecules in shallow aquatic systems (ponds, littoral environments) would have constituted a favourable bed for chemical interactions among molecules and the appearance of new molecular species such as peptides, consisting of an assemblage of several amino acids.

A number of experiments have confirmed the possibility of synthesizing organic constituents from the components present in the primitive atmosphere. Laboratory chemists have tried synthesizing the different families of key molecules: nucleic acids such as RNA, proteins that orchestrate chemical reactions, and phospholipids, which ensure the cohesion of cellular structures in water. In 1953, Stanley Miller produced amino acids by sending electric sparks through a mixture of methane, hydrogen, ammonia and water vapour. One need only bring a sufficient amount of energy to bear upon a mixture containing atoms of carbon, nitrogen, oxygen and hydrogen in order to fabricate some of the elementary 'building blocks' of living beings. The discovery of hydrothermal sources has given rise to new ideas: primaeval organic molecules may have been formed by the reduction of carbon dioxide near underwater hydrothermal sources, as a result of a reaction between hydrogen sulphide (H_2S) and ferric sulphide (FeS).

But once we have the building blocks, how to construct the edifice? Current research indicates that certain organic molecules will spontaneously organize to form vesicles in water. As has been demonstrated in the laboratory, it is even possible to synthesize vesicles resembling cell membranes from molecules of the terpene family (hydrocarbons) without the presence of life. Could this capacity for self-organization explain the appearance of 'proto-cells'? The formation of membranes and the appearance of cells remain among the great mysteries of science.

The theory of *panspermia*, first advanced toward the end of the 19th century, argues that life had extraterrestrial origins. The chemical processes important for the origin of life could just as well take place in outer space, where, according to recent studies by NASA, the molecules necessary for the creation of cell membranes are also present. Thus, considerable quantities of organic precursors might perhaps have descended upon Earth with comets and meteorites: particles of Halley's comet, for example, contain 14 percent organic carbon, and the Murchinson meteorite has been found to contain eight of the 20 amino acids which make up the proteins of life. The cosmic dust that constantly falls on our planet could have transported the same organic molecules as the meteorites. Given the chemistry of the interstellar environment and the discovery of planets beyond our solar system, it is possible to imagine that life may exist or may have existed on other heavenly bodies with a supply of liquid water. These questions merit serious and objective consideration.

From the world of RNA to the world of DNA

In the living world as we know it, proteins are necessary for the production and replication of DNA molecules; on the other hand, it is the genetic message borne by the DNA that determines the sequences of the proteins. It's like the paradox of the chicken and the egg: biologists ask which came first.

In the early 1980s, with the discovery of RNA molecules (ribozymes) capable of catalysing *in vivo* chemical reactions in themselves and other types of molecules, it became possible to imagine primitive RNA molecules capable of transmitting genetic information and catalysing their own replication in the absence of proteins. Such RNAs would themselves have assumed the functions that are now accomplished by DNA and proteins. This gives rise to the hypothesis that a primitive 'RNA world' may once have existed. Recent investigations have confirmed that RNA is responsible for assembling amino acids and proteins in ribosomes. This would mean that ribosomal RNA is an RNA enzyme, a vestige of the primordial world.

It seems very likely that RNA preceded DNA in the course of evolution. A strong argument in support of this hypothesis is that the precursors of RNA, the ribonucleotides, transform into the precursors of DNA, the deoxyribonucleotides, in a complex chemical reaction triggered by advanced enzyme proteins called ribonucleotide-reductases. Thus, to a certain extent, DNA can be seen as a modified form of RNA specialized in preserving information. The DNA molecule is chemically more stable than the RNA molecule, so the genetic message is more faithfully preserved.

3.2 Origins and Dynamics of Intra- and Interspecific Genetic Diversity

3.2.1 Genetic diversity and the universality of life

The present diversity of forms and structures among living things is the expression of a genetic and molecular diversity that is specific to each individual, each population and each species. Understanding the mechanisms that create genetic diversity and the historic processes that enable its perpetuation or, conversely, contribute to its extinction, is important not only to biologists seeking to understand and depict the development of living things, but also for professionals working to maintain and improve the adaptive and productive potential of 'useful' species for their own benefit (conservation and valuation of genetic resources).

The laws that govern processes of mutation and transmission are universal, like the genetic code itself. Their temporal dynamics bring into play chance events, processes of control and environmental filters. As a result, observation of one individual or group of individuals provides no immediate indication of their full genetic potential (genotype). Depending upon the circumstances, the same genotype can express itself in different ways, and conversely, individuals of the same appearance (phenotype) may have different genotypes. The correct interpretation of straightforward observations demands rigorous analysis of the respective roles played by genetic determinism and external factors.

3.2.2 Mutation

3.2.2.1 The phenomenon and its product

The primary producer of diversity is mutation. Any kind of physical/ chemical occurrence that modifies the structure of a DNA molecule alters the genetic information carried by that molecule. The term mutation is used to designate both the process and the product of the change. The modified, mutated DNA carries new genetic information, different from that previously contained.

Through the study of genetics and, more recently, molecular biology, scientists have discovered and analysed a enormous palette of mutational changes that may affect a DNA molecule: chemical modification of a nucleotide; addition or deletion of one or more pairs of nucleotides; addition or deletion of long sequences spanning the size of one or more genes; etc. Such mutations can be lethal, in which case the story ends there. If, however, the individual with the mutations proves viable, the mutations can be transmitted to its descendents and translate, for example, into physiological, morphological or behavioural modifications.

Whatever the chemical mechanisms involved, processes of mutation introduce diversity into a system where the identical reproduction of hereditary patrimony would normally be maintained through the faithful replication of DNA by cellular division. Mutations spontaneously create new alleles. The range of genetic diversity observable at any given time depends, firstly, on the rate at which mutations appear, and secondly, on the possibilities for the mutated forms to develop and compete successfully with other forms of life.

3.2.2.2 The mechanisms of mutation

Mutations can be produced by two different processes:

- by a direct chemical modification of a DNA molecule that alters its information content (modification of a base, etc.);

- during the replication and/or recombination of the DNA, localized secondary structures may form on either strand and interfere with the process of duplication and pairing, causing repetitions, deletions, etc.

Both kinds of occurrences intervene spontaneously but rarely. Molecular processes called mutagens increase the probability of such mutations. Mutagens are the object of intensive study in connection with their ability to produce new genetic information, a capacity that industry is exploiting as a means of diversifying useful micro-organisms. At the same time, the presence of mutagens in the environment is a matter of concern, due to the deleterious or lethal nature of many of the mutations they induce.

3.2.2.3 Systems of repair

When DNA has been subjected to a modification, the cell usually activates its repair systems to correct the change. In the course of its evolution, every species accumulates a whole arsenal of enzymes capable of acting upon mutated DNA, and each individual has its own protective panoply. The repair operation (photo-restoration of thymine dimers, depurination, ligation of broken DNA . . .) may be successful and reestablish the previously existing information. Or else it may be unsuccessful, in which case new information, a mutation, may appear.

These enzymes tend to operate in systems rather than in isolation. The first step is to search out localized faulty DNA pairings (sites that do not comply with the pairing rules A–T and G–C). As a second step, they proceed to produce two corrected DNA molecules (cf. section 3.5.2)

Consistent replication of genetic information is clearly the foundation for the reproduction of cells, individuals and species. But *in vitro* study of DNA polymers shows transgressions of the A–T and G–C pairing rules occurring with a frequency of 10^{-4} to 10^{-5}. By comparison, the *in vivo* frequency of such spontaneous mutations lies well below 10^{-8} or 10^{-9}. This discrepancy is due to the existence of systems for repair. Since each taxon has its own kit of repair systems and its own reproductive system, intra- and interspecific genetic diversity of living things evolve along a multiplicity of diverse routes, involving different qualitative, quantitative and temporal factors.

3.2.3   Variation and stability

At any moment in time, each organism is in a balance between its identical replication and the genesis of diversity through mutation. The former enables clonal multiplication of cells, a process by which, for example, bacterial colonies or metazoan tissues are formed. The latter process constantly produces new information. The flux of genetic innovations is modest, and over a limited period of observation (1–10 years), populations generally appear to retain their own identities. Nevertheless, the living world is by nature unstable; moreover, it is continuously subject to external environmental influences that may also trigger mutational processes. However infrequent, mutations have acted as the motor of evolution and have maintained life on the planet despite its transformations.

Direct impacts upon DNA (radiation, chemical mutagens . . .) have been recognized for some time. They are not specific to the genes that they affect. However, certain environmental influences may exert a more subtle influence by acting upon the structure and activity of repair enzymes or upon chromatin, thereby rendering a particular gene more or less accessible to the repair system. . . . Thus, while the chemical mechanisms of mutation and repair are not correlated to the nature of the genetic information upon which they operate, the probability of their occurrence may nevertheless vary from one individual to another and from one species to another.

3.2.4   Hidden genetic diversity and phenotypic identity

Since the genetic diversity of DNA does not manifest itself systematically at the level of the individual phenotype, two forms of life may appear identical to the observer despite their being genetically very different. *A priori*, it is usually difficult to spot the genes bearing an allelic diversity and to identify the molecular mechanisms that maintain it. However, this is a reserve of hidden genetic diversity that allows for selection in response to exterior change. Farmers, animal breeders and plant cultivators have exploited this property both implicitly and deliberately. By refining their methods of selection, they have been able to extract new varieties from stocks previously considered to be homogeneous.

If practised blindly, selection results in a loss of allelic diversity, as genes of no immediately apparent interest to the selector are eliminated.

As our knowledge of hidden diversity and its applications grows, the importance of genetic resource management is becoming increasingly apparent.

Geneticists study the dynamics of hidden genetic diversity by examining the evolution of DNA sequences. The case of microsatellites is especially illustrative: repetitions of short nucleotide motifs (CA, for example) are dispersed in the DNA. When they are situated in the non-coding zones of the genome, the number n of repetitions has no effect upon the phenotype. Observation shows that n may frequently vary from one individual to another. Their mutability (around 10^{-2}) is so high that microsatellites are used to expose the genetic diversity between related individuals and establish their lineage. Microsatellites also exist in the coding zones of genes. Here, the possible degree of variation is limited: the motifs must be repeated in triplets to maintain the legibility of the genetic code, and their number is restricted by the extensions or contractions of the primary structure of the protein that are compatible with its function. In this case, changes in n often have distinctive phenotypic effects upon individuals. The rate of mutation may be so high that n changes from one generation to the next. For certain hereditary diseases in humans (Huntington's disease, for example), the severity of the pathology increases in families where n increases.

3.2.5 Spatial organization and dynamics of intraspecific genetic diversity

Genetic diversity is the fruit of the history of the DNA molecules present on the planet today. It is distributed geographically over all the species inhabiting the different ecosystems. The developmental and reproductive strategies of each species, population and individual mould this genetic diversity both qualitatively, quantitatively and over time.

Theoretically, in the absence of any intrinsic or extrinsic constraints, all genes and all DNA molecules are equally prone to replication. Under such conditions, genetic diversity is preserved from one generation to another and, in the absence of mutations, remains identical onto itself. The pattern of diversity observed in a site remains stable, characterized by the frequency of the various alleles present in the area. Thus, for sexually reproductive plant and animal species, mitosis and meiosis are dependable and accurate systems for splitting DNA molecules, independently of the nature and volume of information carried by these molecules.

The laws of genetics revolutionized biology in the 20th century. They are quantitative and predictive and may be tested experimentally by comparing actual observations with the theoretical predictions of a mathematical model. This applies to Mendel's laws on the scale of individuals, as well as to the Hardy–Weinberg law on the scale of populations. (The latter law predicts that for any population, the nature and frequency of genotypes will remain constant from one generation to the next, provided all encounters between gametes of the previous generation are equally probable – a situation known as *panmixia*.)

Given a large population, its genetic diversity should be preserved over successive generations. However, since associations between gametes are subject to chance encounters, it is likely that certain alleles will not participate in the reproductive process, and therefore, the genetic diversity of the population will decrease over time, even if the probability is low. This effect is known as *genetic drift*.

In practice, the depletion of diversity predicted by this theory is strongly influenced by the size of the population (the risk is greater in small populations) and by its demographic characteristics (temporarily high mortality and reduced size – constrictive bottlenecks, overlapping generations, consanguinity, fragmentation of the population as a result of the appearance of barriers or parasites, emigration, etc.).

Conversely, the immigration of genetically different individuals or the appearance of new mutants are phenomena that enrich the genetic diversity of populations. Like panmixia, genetic drift can be mathematically formulated and modelled to test the role of these processes within the dynamics of genetic diversity observed in a given place over a given time and for a given population (or species).

In reality, mutations, genetic drift and random demographic circumstances are not the only factors to have shaped the genetic diversity present on the planet. Today, we are only familiar with the individuals who succeeded in reproducing and developing, while in fact they were once in competition with others who inhabited the territory at the same time but disappeared. It was Darwin, in the 19th century, who discovered the essential role of natural selection in the process of evolution.

Mutations are the motor of evolution. They provide the material upon which drift and selection subsequently act. Selection accounts for the difference between the evolution of genetic diversity in reality and such as it would be if a population were subject only to mutation and drift. The selective value (or fitness) of an individual reflects how successfully its genes are transmitted from one generation to the next. There are working theories on how to model genetic diversity under these conditions. They make it possible to compare different populations and species and predict their evolution with respect to the disturbances affecting them.

3.3   How are Species Born?

The formation of a new species, or *speciation*, results from one of the following two scenarios:

- the replacement of one species by another, following an accumulation of adaptive genetic transformations over a period of time. This is known as speciation by *anagenesis*;

- the emergence of two or more species deriving from a pre-existing species, whose populations may, for example, have been separated geographically from one another. This is known as speciation by *cladogenesis*.

Cladogenesis and anagenesis combine in the evolutionary tree: cladogenesis explains the diversification of life, while anagenesis explains its continuity.

3.3.1   Mechanisms of speciation

The basic mechanisms for the speciation of living things operate in the realm of genetic heritage. Three processes are at work in the dynamics of intraspecific diversification: mutation, selection and drift. *Ipso facto*, they generate the diversity from which new species emerge.

3.3.2   Modes of speciation

Amongst the factors and mechanisms that enable the process of speciation, two mutually non-exclusive scenarios are as follows.

- The classic model is that of *allopatric* speciation: populations of the same species become geographically isolated by events such as the separation of continents, the appearance of protuberances constituting

a barrier to exchange, the isolation of watersheds, etc. The populations evolve independently by processes of mutation, selection and
drift, to the point where, after a certain amount of time has elapsed,
they are no long able to reproduce with each other. Thus, they have
given birth to new species.

- Over the last decade, scientists have gathered a considerable amount
  of evidence for a kind of speciation called *sympatric* (speciation in
  'one place' without geographic isolation); i.e., new forms are capable
  of isolating themselves sexually through mutation, selection and drift,
  even while coexisting in the same ecosystem. It seems that divergences
  appear within populations as they become specialized in the utilisation of certain resources and that these divergences amplify to the
  point where new species are born. Thus, it is very likely that the
  adaptive radiation of the cichlids of the east African lakes (see
  below) is the result of sympatric speciation. In the small crater lakes
  of Cameroon, there is evidence that the existence of several endemic
  species of cichlid fish was the result of sympatric speciation combined
  with a diversification of trophic behaviours.

Both kinds of speciation are linked to the occupation of new niches for
which the species were not adapted *a priori*. But in the case of allopatric
speciation, external developments impact biological processes, often over
a long period of time, for example by creating new opportunities and
habitats. In sympatric speciation, on the other hand, certain individuals
belonging to the same species and sharing the same geographic space
acquire new functions or differentiated means of using their potential.
The underlying mechanisms are not identical. The first case involves a
coevolution of species and milieu through a process of trial-and-error:
the apparition of evolutionary novelties and natural selection. In the
second case, the motors of evolution reside within the biology of the
species itself.

3.3.3 Gradualism and/or punctuated equilibriums

According to the theory of gradualism, species transform progressively
over time, proceeding from a mother species to a daughter species. Slight
changes accumulate over great lapses in time, leading to gradual differentiation. This theory is very popular and is central to Darwin's

thoughts. Transformations on the microevolutionary level culminate in speciation; macroevolution is nothing more than the large-scale description of the consequences of microevolutionary mechanisms.

While the theory of gradualism applies neatly to situations of sympatric speciation by cladogenesis, for example, it does not explain the appearance of major and radical modifications on a macroevolutionary level, for example, in the organizational plans of organisms. Based on a study of a series of fossils from the Secondary and Tertiary eras, Eldredge and Gould suggest that these may be due to sporadic occurrences of sudden changes: long periods of stability punctuated by phases of rapid evolution. This model favours the idea of allopatric speciations operating over short intervals of time, after which the species remain stable. This version has still not been formally substantiated and is the object of controversy.

3.4   Extinctions

Throughout the history of the Earth, species have emerged, while others have disappeared. In other words, a species is born, lives and dies. Extinction is a normal process of evolution. The history of life over almost 4 billion years has been studded with periods of crisis, when large numbers of species disappeared. Many lines of animals and plants have become extinct, and the entire biological diversity of the present time amounts to a mere 1% of all the species that have lived in the past.

Palaeontological archives provide evidence for five major events that have left their mark upon the last 500 million years (Figure 3.1).

- 440 million years ago, at the end of the Ordovician period, a major event must have caused the disappearance of 85% of all species. The trilobites, cephalopods, brachiopods and echinoderms were severely affected. However, no large phylum appears to have disappeared entirely.

- 365 million years ago, the extinctions of the Devonian period eliminated around 75% of the marine species. Most tribolites disappeared. This event profoundly upset the reef systems, whilst continental plants and arthropods continued to evolve undeterred.

- The extinction event of the Permian period, at the end of the Primary era (245 million years ago), was the most severe of all. Certain

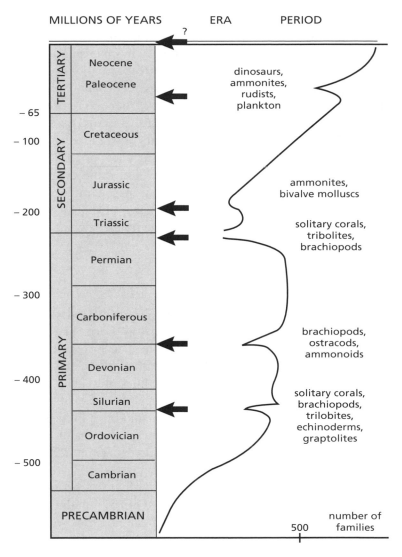

Figure 3.1   Times of the different geologic eras and principal mass extinctions, based on palaeontological records

Palaeozoic groups vanished entirely. Close to 95 per cent of all marine species disappeared, including the last of the trilobites and graptolites, as well as the benthic foraminifers and certain corals. The event also impacted continental environments, where it wiped out two-thirds of the insect families and 70 per cent of the families of vertebrates. The cause for this wave of extinctions is unknown, but it could have been the result of a major shift in climate.

- The event at the end of the Triassic period began 215 million years ago and lasted for about 15 million years. It likewise brought about the extinction of 75 per cent of the marine species of the time, as well as some terrestrial species.

- The event occurring at the end of the Cretaceous period (K-T extinction) 65 million years ago is surely the most famous, since it marked the demise of the dinosaurs. The ammonites, as well as many marine foraminifers also vanished at this time. Plankton and marine benthos were severely diminished, as was a large proportion of terrestrial vegetation.

There is an ongoing debate as to the magnitude of these crises and whether they arrived abruptly or progressed gradually. Among the palaeontologists and geologists trying to determine the causes of these mass extinctions, there are some who favour catastrophic explanations: a unique event with consequences for the entire planet, such as the impact of an asteroid, eruption of a volcano, could have been responsible for a series of chain reactions, from the collapse of certain ecosystems to the resulting extinctions. Others invoke geological upheavals, the mutational effect of cosmic rays, shifts in ocean currents, resulting in continental collisions, and, of course, climate changes. Quite likely, major crises resulted from the coincidence of several causes relating to the history of the global environment.

Did these momentous events in the history of the Earth have a significant impact upon biological evolution? Some scientists hold that mass extinctions reshuffled the cards of the evolutionary deck. Even if it is difficult to evaluate their impact, there is no doubt that these great crises modified the course of evolution, since they gave the living world the opportunity to reorganize itself. These periods of crisis constituted ruptures in the evolutionary process that may have helped new biological forms to become established; however, they were probably not the actual force behind their appearance. Finding themselves less subject to competitive pressure than before, the surviving organisms were able to recolonise environments that had become more hospitable as a result of the crisis and give birth to new diversifications. There are some who go so far as to assert that we have entered a sixth period of mass extinction, for

---

Cambrian explosion

The Cambrian explosion gave birth to many original forms of organization, most of which fell prey to later extinctions and disappeared
again. Evolution has proceeded on the basis of successive decimations.
Surviving lines have recolonized the ecological niches relinquished by
the deceased and differentiated into new species, families, classes . . . ,
which have, in turn, been partially decimated by later catastrophic
episodes. Overall, the diversity of evolutionary branches (i.e. of the
organizational charts of life) has diminished from the Cambrian age to
the present, while the global number of species has grown over the
course of the geologic ages. This perspective reveals the importance of
extinctions, whilst underlining their fortuitous nature.

---

which humankind, with its galloping demographics and global activities,
is responsible.

3.5   Adaptation: a Fundamental Property of Living Organisms

Biological diversity is dynamic by nature. The physical and biotic environment of a species is continuously changing: climates are variable,
competitors invade the domain, sources of nutrition change, etc. Within
limits, organisms are capable of adapting to changes in the world within
which they evolve. Species will always undergo modification in the long
term, whether on the genetic, the biological or the behavioural level;
some are born, and others die. Adaptation is a functional mechanism
that enables species to deal with variability in the conditions of their
living quarters. This capacity on the part of living beings is essential to
their survival.

  The term 'biological adaptation' covers many facets: the genetic aptitude of living beings to adjust to different and changing circumstances;
the manifestation of this aptitude at any given moment; the mechanisms
at work; and the notice that society takes of these different manifestations. Actually, there are two basic types of adaptation:

- the adaptation of each individual, as an expression of his capacity to live and succeed in a changing context;

- the conquest of new environments to sustain different individuals, the fruits of biological evolution.

3.5.1 Genetic diversity and adaptation of eukaryotes to environmental changes

Biological diversity, as we know it today, might appear to provide an optimized response for every situation in the sense that each species seems to make the best of its available resources. Is this really true, and what are the processes that have made it possible to attain such a state?

3.5.1.1 Individual adaptation: phenotypic plasticity

Adaptation is based upon the combined use of different aptitudes. At the molecular level, proteins, enzymes and subcellular membrane walls are restructured on a very short-term basis, thereby modulating their catalyst or transportation activities. The environment also intervenes over the long term (the duration of an individual life span, for example), by regulating the expression of the genes that control these molecular actors and influencing such activities as cellular differentiation and proliferation, morphogenetic and physiological plasticity, behaviour and reproduction.

Adaptation may involve modification of a phenotype in response to a specific signal received from its environment. The modification improves biological functions such as growth, reproduction and/or survival. Thus, on the local level, adaptation can be characterised as increasing the frequency of those biological traits that promote survival or greater reproductive success within the population in question, under the particular conditions provided by its environment.

A basic key to adaptation lies in the plasticity of organisms. The biological characteristics of a phenotype – whether morphological, physiological, behavioural, etc. – are actually the result of an interaction between its genes and the environment. It must exercise flexibility to be able to exploit the resources necessary for the survival of the species. Phenotypic plasticity corresponds to the diversity of adaptive responses produced by a single genome. These responses may then be subjected to natural selection. The basic rule is: the phenotype is the product of the genotype and the environment, but it is also the object of selection!

Behaviour is another phenotypic characteristic that demonstrates great plasticity. For some biologists, the expression of a phenotype first manifests itself through a change in behaviour, particularly when a population has to come to terms with a new habitat or niche. A change in behaviour is almost always the first step towards evolutionary change. Amongst the hypotheses that focus on the interactions between evolution and changes in animal behaviour, a major role is played by the capacity for learning that enables animals to exploit new situations and gain access to new resources. Imitation and learning facilitate the assimilation of 'evolutionary novelties' and their transmission within the populations involved.

3.5.1.2   Collective adaptation: natural selection

Thanks to genetic polymorphism (each individual member of a species is genetically slightly different from the others), the individuals constituting a population are able to respond to environmental constraints in slightly different ways. Chance generates variations (mutations), that are then strained through the sieve of selection. This is what Darwin termed *natural selection*, as opposed to the artificial selection practised by breeders. The principle of natural selection implies two complementary processes: the existence of an hereditary genetic variability, and a phenomenon whereby the best performing individuals of a given environment are selected at the reproductive level. When conditions change, those genotypes that produce the phenotypes best suited to respond to the new constraints enjoy an adaptive advantage and are selected in the course of successive generations. Selection essentially affects the frequency of genes: genes controlling adaptations that reinforce the chances of successful reproduction will be favoured, and their frequency can thus grow over successive generations.

The selected genes may tow along other genes which are not themselves selected, but whose expression is favoured indirectly in this way. Thus, the theory predicts that alongside of progressive transformations, large-scale qualitative leaps may occur when unexpected mutations reshuffle the deck of cards in the game of selection and its consequences. Seen from a different angle, variations in the structure and function of ecosystems are both the cause and effect of natural selection. In this regard, scientists speak of the 'cycle of biological diversity': mutations and the biophysical environment are the sources of variability; the product of adaptation and selection is biological diversity; which, in turn, interacts with and modifies the environment.

3.5.1.3   Creative adaptation: ecological innovations

Under what conditions will a species invade a new ecological niche, a different environment to which it is not adapted? And how do new functions appear in individual members of this species? These questions are fundamental to understanding the mechanisms of evolution in a perpetually changing environment.

The evolutionary process known as *adaptive radiation* involves the colonization of different ecological niches in the same ecosystem by populations or species descended from a common ancestor. The term 'species flocks' is used in this connection to designate groups of endemic species that are morphologically very close. They are descended from an ancestral species whose populations have progressively differentiated by specializing their use of the different resources (nutrition, habitat, breeding grounds, etc.) provided by their environment. This would seem to be a way of optimising the use of disposable resources. At the same time, this process translates into changes in the general functioning of the system, further complicated by the flux and flow of matter and energy in many directions.

A well-known example of adaptive radiation is the case of the finches studied by Darwin on the Galapagos Islands. The 13 species that have been identified are all considered to be descended from one common ancestor, who reached the Islands several million years ago. Each species settled on an island, in a particular type of habitat, and then proceeded to differentiate from the others. The whole set of species issuing forth from this process is better adapted to using the different local resources than its single ancestral species.

3.5.2   The adaptive capacity of prokaryotes

The key to the evolutionary success of micro-organisms lies in their extremely brief generation cycle (about one hour for some bacteria) and their capacity for adaptation. Under given circumstances, bacteria are capable of exchanging genes, tantamount to a primitive form of

---

Adaptive radiation in lakes

A famous example of adaptive radiation is that of the cichlid fish of great lakes of east Africa. The 'haplochromins' of Lake Victoria are descended from a single ancestral species of river fish that must have colonised the lake and given birth to some 300 living species currently occupying a large variety of ecological niches. The cichlids have developed trophic specializations to the point where they make use of all available resources, including some that are used by them alone. In addition to trophic specialization, they also exhibit specialization in their reproductive behaviour, including mating rites, specific sexual colouration, and territorial and tribal behaviour.

Similar phenomena have been observed in other African lakes, such as Lake Malawi and Lake Tanganyika, as well as in some so-called 'ancient' lakes (older than 100 000 years), such as Lake Titicaca, Lake Baïkal, etc.

---

sexuality. Enabling the incorporation of exogenous DNA, this is a highly effective process of adaptation. Bacteria employ different mechanisms to accomplish this feat.

- The first process, *conjugation*, is probably responsible for most of the transfers. Plasmids, DNA molecules in found cytoplasm that are independent of chromosomes, encode proteins, enabling one bacterium to attach itself to another. Plasmids can also pass from one bacterium to another, bearing both their own genetic information and possibly part of the chromosome from the first bacteria. These DNA molecules may recombine with the chromosome of their host cell, thereby generating genetic diversity.

- Another highly effective transfer mechanism is *transduction*. In this case, the vector is a bacterial virus (bacteriophage) that passes from one bacterium to another, bearing its own information and sometimes also a piece of the chromosome from the first bacterium.

- *Transformation* is the ability for the cell to be penetrated by a free DNA molecule encountered in the milieu of the bacteria.

Studying the flux of genes among bacteria has become a field of investigation in its own right. It is teaching us more about the relationships between species and the capacity of prokaryotes to adapt to environmental change. Much of this is still uncharted territory, particularly in the realm of soil bacteria.

Bacteria may undergo yet another adaptive process that exploits imperfections in the system for maintaining the fidelity of DNA replication. Bacteria have mechanisms to prevent and repair modifications produced by mutation, especially when DNA is replicated (see section 3.2.2.3).

- The *system for repairing mismatches* (SRM) functions by activating a group of proteins that intervenes when the rules for pairing the constitutive elements of DNA are not respected. Errors of this nature may occur during replication. The protein MutS detects such errors and activates other proteins (MutH and MutL) that repair the faulty strands; thus, SRM acts to deter mutagenesis.

- The primary function of the SOS system is also to repair DNA, but it is the reverse of the SRM system. It is triggered in times of stress, when there is a threat to the integrity of the genome, or when the mechanism for replication is impeded and single-strand DNA is produced. The SOS system stimulates mutagenesis, increasing genetic variability, and causes a rearrangement of chromosomes, sometimes even to the point of integrating DNA fragments from another species. This encourages bacteria to produce different descendants until some appear optimally adapted to the new environment. Once the stress period is over, the SOS system is deactivated. The process of replication resumes its course, re-establishing a normal rate of mutation and once again giving priority to the mechanisms that preserve the stability of the genome (such as SRM, for example).

Microbes, the champions of adaptation

Microbes are particularly quick to adapt to environmental change. In every microbial population, there are individuals called mutators that consistently produce a great degree of variability among their descendents. This variability is usually of no use, but it assumes adaptive value when there is a sudden and severe change in environmental conditions: among the many variations, there may be individuals

continues overleaf

Microbes, the champions of adaptation (*continued*)

with advantageous mutations that are better adapted than the initial strain and end up replacing it. Nevertheless, bacterial strains tend to adapt by genetic transfer between bacteria more often than by mutation. There are different examples for this natural genetic ingenuity which enables bacteria to adapt rapidly to changes in their surroundings. Soil under cultivation to grow corn adapts to the use of atrazine within a couple of years. Today's cornfields all contain bacteria capable of disintegrating this herbicide. Another example is the soy bean, a Chinese vegetable introduced to North America without its symbiotic, nitrogen-fixing bacteria. Within a few decades, some American bacteria developed a symbiotic relationship with the soy bean with the same effectiveness as their counterparts on the continent where the plant originated.

3.6   Major Stages in the Diversification of the Living World

From primitive soup to the first organisms, from single-cell to multi-celled organisms, from aquatic to terrestrial environments, life has grown more complex over time.

3.6.1   The main lines of evolution and how they are related

The tree of life, as deduced from studying the morphology of living species and fossils, has been called into question by recent research. Traditional schemes postulated a simple dichotomy in the living world between eukaryotes and prokaryotes; however, one of the most spectacular yields of molecular phylogeny has been to demonstrate the existence of three superior categories above the kingdom level: the eukaryotes (Eukarya), eubacteria (Bacteria) and archaeobacteria (Archaea), with the last two groups constituting the prokaryotes.

Prokaryotes appeared about 3.5 billion years ago. Since this time, they have adapted and prospered on a planet whose environmental characteristics have been continuously changing. They are present in almost all environments where life is capable of existing and have

Bacteria

A bacterium (from the Greek *baktêrion*, meaning stick) is a cell surrounded by a membrane and containing all the elements necessary for its own reproduction. In this respect, bacteria differ from viruses, which are about one-tenth the size of bacteria and need to invade a cell in order to reproduce. Bacteria are the most abundant living organisms on Earth: despite their miniscule size (on the order of about one-thousandth of a millimetre), their cumulative mass would be comparable to that of all the plants on Earth.

in their turn contributed towards modifying the terrestrial environment. Compared to the eukaryotes, prokaryotes do not have a real nucleus enclosed by a membrane; instead the DNA forms a tangle called nucleoid. This tangle actually corresponds to a single bacterial chromosome whose DNA molecule carries all the genes necessary for cellular life.

The prokaryote cell may comprise other, smaller DNA called plasmids with only a few genes each. These plasmids reproduce independently of the principal chromosome, and many of them are capable of switching to other cells while the bacteria are conjugating (see Section 3.4.2). Prokaryotes reproduce asexually through a form of cell division called scissiparity. In a favourable environment, a bacterium can go through repeated divisions and give birth to clones of identical cells, increasing exponentially in number.

The discovery of archaeobacteria in 1977 was tantamount to a scientific revolution. These organisms were first detected in extreme environments (hydrothermal sources on the ocean floor, highly saline waters, or acidic environments) that may resemble some habitats of primaeval Earth; but they are actually present in almost all habitats. New technologies of molecular biology have made it possible to establish their presence in marine plankton, soils and continental freshwater bodies, although they have never yet been detected in *in vitro* cultures.

Eukaryotes are unicellular (protist) or multicellular organisms constituted by an aggregation of specialised tissue (animal, plant, fungal). First appearing around 1.8 billion years ago, they are characterised by a cellular nucleus surrounded by a double membrane and containing genetic matter organized in chromosomes. The cell also contains other well-defined organelles such as mitochondria, lysosomes, etc.

How did the transition from prokaryotic to eukaryotic cells occur? As yet, little is known about the mechanisms involved. Mitochondria and plasts, which are organelles present in eukaryotes, have a genome and independent enzyme devices for synthesizing proteins. Their characteristics closely resemble those of bacteria. According to the endosymbiotic hypothesis, a primitive eukaryote could have phagocytosed a free bacterium, and the two cells could have embarked upon a lasting symbiotic relationship. Gradually, the phagocytosed bacteria would have lost their capacity to live independently. Thus, the chloroplasts would appear to be descended from cyanobacteria, and mitochondria from purple bacteria.

3.6.2   From unicellular to multicellular organisms

An important transition took place from single-cell eukaryotes such as algae and protozoa to multicellular organisms, or metazoa, with specialized cells for tissue formation, absorption of nutrients, respiration, reproduction, etc. This transition is believed to have occurred around 1 billion years ago. However, the first multicellular organisms were probably microscopic in size and have left us no fossil traces. Multicellular algae have been identified in sediments from Spitzberg dating back to 800 million years ago. The earliest known macroscopic metazoa are pre-Cambrian fossils such as the fauna of Ediacara (a site in southern Australia): flat, soft creatures without mineralized skeletons, resembling worms that lived around 600–550 million years ago. This mysterious fauna resembles none of the organizational types existing today, and it is still unclear how they might be related.

3.6.3   The explosion of biological diversity in the Cambrian period

The major groups of animals, many of which have survived to our day, appeared quite suddenly and almost simultaneously at the beginning of the Cambrian. This is sometimes called the *Cambrian explosion*, because of its extraordinary burgeoning of animal life. The Burgess fauna of Columbia, popularized by J. Gould, dates back to slightly over 500 million years ago. It contains worms, molluscs, various arthropods and a chordate (*Pikaia*) which may have been the ancestor of the vertebrates. In addition to those forms related to present-day forms, it also contains unclassifiable forms, some of which probably belong to extinct branches.

The Burgess fauna also teaches us about the apparition of heterotrophy and the first communities linked by food chains. This is the epoch of the first predators.

Comparing all these organisms, it is clear that anatomical organization was far more diverse than it is today. Various explanations for this phenomenon have been advanced:

- multiple genomic combinations were possible, because the genomes of multicellular animals were less complex than they are today;

- there were still many unoccupied ecological niches – a situation which is favourable to evolutionary innovation.

Thus, the history of life is written by the success of certain forms of organization, emerging from the far richer stock initially created in the Cambrian explosion (see Figure 3.2).

3.6.4   From the sea to the land: a successful transition

To colonize new habitats when they emerge and adapt to the new conditions that these present is a fundamental characteristic of life. The conquest of dry land by living organisms falls into this category, since firm ground emerged only gradually from the oceans that covered most of the planet. The first forms of life to colonize the continents were probably cyanobacteria resistant to ultraviolet light. Still, fossils show that life remained essentially aquatic throughout the Cambrian. The earliest forms of terrestrial vegetation were the bryophytes, non-vascular plants derived from green algae (represented today mainly by mosses). The subsequent colonization of dry soil by plants is a remarkable case of evolutionary adaptation. Terrestrial plants could only survive and grow because they acquired structures and mechanisms adapted to the new living environment, such as:

- a waxy pellicle called the cuticle that prevents plants from drying out, together with specialized cellular devices called stomata, allowing gaseous exchange with the atmosphere through the watertight cuticle;

- a vascular system for transporting the water and nutritive salts extracted from the soil by the roots;

MYA   ERA   PERIOD

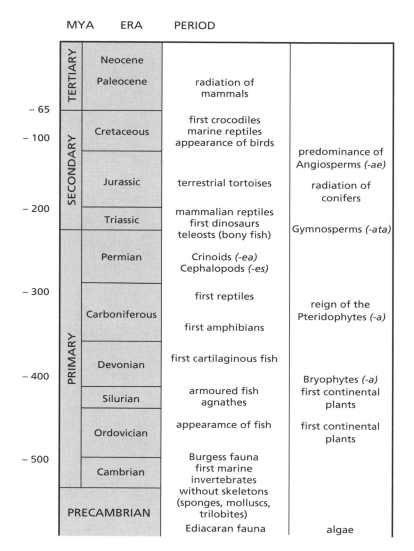

Figure 3.2 Chronological appearance of the principal groups of animals and plants in the history of evolution

- spores, protected by a resistant cover to prevent them from drying out, that can be disseminated by the wind to ensure reproduction.

The pteridophytes (nowadays represented by ferns, equiseta or scouring rushes, lycopodia or club mosses), began to evolve in the Silurian around 450 million years ago. These plants equipped themselves with roots and leaves and sometimes a woody tissue that made their stems more resistant. As their trunks became reinforced by wood and enveloped by tree-like bark, the ferns grew higher and larger. The apogee of the pteridophytes during the Carboniferous period represents the first major radiation of terrestrial vegetation. This period witnessed the spread of the great equatorial forests, composed primarily of giant horsetails and arborescent ferns, contributing to the vast accumulations of carbon in the sediments of this time. Forests populated mainly by large arborescent forms, such as lepidodendrons and sigillairias, evolved in marshy zones.

During the transition from the Primary to the Secondary Era, pteridophyte flora gave way to predominately gymnosperm flora. Spores appeared with the prespermaphytes, represented by *Ginko biloba* today. After fertilization, which requires a tiny amount of water, the spore detaches itself from the plant and develops in the soil. Spermatophyte gymnosperms predominated throughout a good part of the Secondary Era, their flora providing food for herbivorous dinosaurs. Modern descendants of the gymnosperms are conifers, such as firs and pines.

During the Cretaceous period, at the end of the Secondary Era around 100 to 130 million years ago, the appearance of angiosperms, or 'flowering plants' radically altered the composition of flora over all soil surfaces projecting from the seas. Transported by the wind, pollen can travel hundreds of kilometres before fertilizing an oosphere. After fertilization, the flowers transform into fruits that contain and protect the seeds. Dissemination through seeds is thought to represent a form of adaptation to unfavourable conditions such as winter cold. The hegemony of the Gymnosperms declined progressively, and a number of families became extinct. Meanwhile, from the time of their incipience, angiosperm flora began to diversify: all the families known today were represented. This flora probably originated in an equatorial forest, whence it spread over the surface of the earth.

The current distribution of vegetation over the globe derives from flora that took root at the end of the Secondary Era and experienced the effects of the great ice ages occurring at the end of the Tertiary Era (5 million years ago) and throughout much of the current Quaternary Era.

As for animals: arthropods (myriapods, scorpions) began their conquest of terrestrial habitats approximately 400 million years ago, soon followed by collembola, which proliferate in fungi and decomposing matter, and finally insects. External skeletons in the form of a rigid coating (the cuticle) and jointed appendages were already characteristic of primitive aquatic arthropods. With this rigid carapace to protect them from ultraviolet rays and prevent them from drying out, they were predisposed for the conquest of the aerial environment. Gills were replaced by tracheal systems. A decisive innovation for insects was the appearance of wings, probably in the Devonian, enabling them to escape from predators, as well as spread rapidly and colonize new ecological niches.

3.6.5   The long history of vertebrates

Most Cambrian fossils have carapaces that serve as protection. The appearance of an external skeleton was followed by the development of jointed appendages. Other animal groups proceeded to develop a segmented vertebral column serving as an anchor for muscles. The earliest known vertebrate fossils are vestiges of 'fish' discovered in China and dating from 530 million years ago: small cartilaginous animals that lived amongst plankton.

The earliest jawless fish (or agnatha) with a vertebral column appeared around the Silurian Period, 420 million years ago. By the time of the Devonian, they had all but disappeared, with the notable exception of today's lamprey. They gave birth to fish with jaws: the armoured placoderms and the cartilaginous selaciens (represented by skate fish and sharks today). Bony fish diversified in the time from 400 to 350 million years ago. Of the crossopterygians, only one variety remains today: the coelacanth or *Latimeria*.

Deriving from a common ancestor, tetrapod vertebrates developed legs 370 million years ago. Contrary to earlier hypotheses, this had nothing to do with their leaving the water, but appears rather to have been a means of adapting to life in shallow waters encumbered by branches. This does not exclude the probability that their legs would have been progressively modified to accommodate terrestrial locomotion. This would have favoured the first tetrapods who set out to explore the beaches and rocky coastlines – beginning with the amphibians (360 million years ago), followed by the reptiles (300 million years ago) – and eventually colonized terrestrial environments.

Aquatic vertebrates perfected a number of technical innovations to help them to change location and live on dry ground. In particular, they devised adaptive responses to a trivial problem, namely gravity, such as developing new muscles to support the weight of their entrails. However, for a long time, they remained dependent upon aquatic surroundings for purposes of reproduction. The appearance of the *amniotic egg*, with its semipermeable shell enveloping nutritional reserves and protecting the embryo within, constituted a major leap towards independence from the aquatic milieu. This complex system enabled a number of evolutionary branches to develop; among them, the mammalian branch, which eventually abandoned the eggshell and adopted viviparity.

Beginning in the Triassic (240 million years ago), the reptilian group gave rise to the dinosaurs, who disappeared 65 million years ago. These warm-blooded animals belonged to several large groups comprising a great variety of species, some of which were enormous in size. Mammals descended from a line of reptiles toward the end of the Primary Era, around 250 million years ago. The earliest – tiny and humble – representatives of this group were contemporaries of the early dinosaurs. The first true mammals emerged in the Triassic around 200 million years ago and looked somewhat like shrews. The extinction events of the Cretaceous/Tertiary periods caused the great reptiles to disappear, vacating many ecological niches that they had occupied, while furred animals progressively diversified and colonized all available environments.

Humankind: a primate success story?

Humankind belongs to the vast family of primates that appeared in the Eocene, around 65 million years ago, and diversified during the Tertiary period. The divergence that culminated in humankind is written in the history of the order of primates. Hominoid primates emerged around 20 million years ago. The diversity of this group and its geographic expansion testify to its incontestable evolutionary success, although only a very few families have survived into the present. Today's anthropoids have two distinct lines: *Hylobates* (the gibbons of the forests of southeast Asia) and the hominoids, currently represented by orang-outangs (pongids), gorillas, chimpanzees and humans (hominids). The recent discovery of anthropoid primate fossils in Asia implies that the divergence between the Asian and African lines of descent occurred no later than 35 million years ago. The divergence between humans and chimpanzees is charted at around 5 million years ago, but there is a dearth of information on the period from 5 to 14 million years ago. Genetic analyses show that human beings and the great apes of Africa are still closely related today (Figure 3.3).

continues overleaf

Humankind: a primate success story? *(continued)*

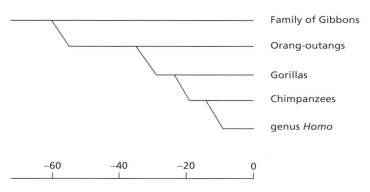

Family of Gibbons

Orang-outangs

Gorillas

Chimpanzees

genus *Homo*

-60   -40   -20   0

Figure 3.3   Evolutionary tree of the group of anthropoid Primates from which *Homo sapiens* emerged

The most ancient known fossils in the human line of descent are *Australopithecus*, dating back 4 million years. Several species coexisted in Africa between 4 and 3 million years ago. Also in Africa are the remains of *Paranthropus*, who lived between 2.5 and 1 million years ago and was closely related to *Australopithecus*, although no ancestor to *Homo sapiens*. They could be regarded as a kind of 'discarded experiment' on the part of evolution. The oldest representatives of the genus *Homo* would have appeared in Africa around 2.5 million years ago and thus been contemporaries of the last of *Paranthropus*. Several species of the genus *Homo* have been identified: *Homo habilis* disappeared around 1.6 million years ago; *Homo ergaster* embarked upon the conquest of Asia and Europe around 1.8 million years ago. Those who colonized Asia became the ancestors of *Homo erectus*; others settled in Europe, giving rise to *Homo neanderthalensis* (Neanderthal man), who lived between 100 000 and 30 000 years ago. Other African populations developed towards ancient *Homo sapiens* between 400 000 and 100 000 years ago. *Homo sapiens* subsequently spread over Europe (Cro-Magnon man) and would have cohabitated with Neanderthals in a number of regions during the period from 40 000 to 30 000 years ago. This scenario is plausible, but it is not supported by all palaeontologists. There are some who hold that evolutions from ancient populations towards modern humans occurred on a parallel basis, with continual intermingling among the different lines. In any case, current hypotheses are likely to be revised in the light of new discoveries.

The history of humans appears to be ramified rather than linear. Various 'human' species appeared and disappeared. For the moment, it is not easy to ascertain to what extent they were interrelated, but clearly these species, some of which were contemporaries, evolved along parallel lines. Of all these evolutionary possibilities, one single form emerged to conquer the planet. Nevertheless, the variety within the group of *sapiens* remains considerable.

3.7   Is the Evolution of Life Predetermined?

Since the very beginning of life on Earth, the living world has been engaged in a perpetual process of transformation. Some species populated the Earth before disappearing with no heirs. Others gave birth to new lines occupying different ecological niches. Overall, the appearance and succession of different groups of living beings appears to be characterized by an increasing degree of complexity. Since life first appeared, the world has witnessed the progressive emergence of beings characterized, in particular, by increasingly complex behaviour. In the case of the vertebrates, despite the various extinctions to which they were subject, the anatomy of the nervous system continued to develop and produce corresponding changes in behaviour. The power of reflective thought is thus the result of a process that has been underway for billions of years.

One fundamental question that biologists ask themselves is whether chance has been the only captain on board during the long voyage through the evolution and diversification of the species or whether there is an 'invisible hand', an unidentified guiding force that lends sense to evolution. To phrase the question differently: does the evolution of the species translate into progress?

The evolution of life over the geological ages is usually perceived as a process regulated by natural selection: preserving 'useful' variations and eliminating or ignoring those that are of no use. Although Darwin rejected the concept of evolutionary progress as a law compelling organisms towards increasing complexity, he nevertheless accepted the idea that natural selection contributed to the improvement of organisms with respect to their living conditions.

Evolution is often simply represented as a tree with an increasing number of branches, some of which are withered away. But the palaeontological facts tell another story. Organizational diversity was greater during the Cambrian than today. Moreover, after the great catastrophes, only a small number of species survived to diversify anew.

For many authors, chance is the only explanation for the evolution of life. As they see it, the only object of the evolutionary game is to select those organisms that will yield the most descendents, and the only 'reward' is that the game can be continued if the process is successful. There doesn't seem to be any long-term global perspective. If this is the case, ecologists may be asked the following question: if evolution is simply a question of chance without any apparent form of determinism, what is the sense of studying the natural order and seeking evidence for general laws?

To summarize this debate as it stands today.

1. The course of evolution is punctuated by periods of crisis. The species that survive these crisis are spared, not because they perform better in line with the theory of natural selection, but rather because at the moment of the event responsible for the mass extinction, they have the luck to possess a biological trait favourable to their survival. Such biological traits are developed *a priori* and independently of the factors responsible for the mass extinction.

   Mass extinctions preserve or eliminate species by chance. Survival or disappearance is above all a question of good or bad luck with respect to chance events that are not predictable on the basis of present knowledge. There is no way to distinguish the survivors from the victims. The living world of today is the result of a cascade of events occurring since the time immemorial, extending well beyond the realm of intervention by selection. If one could rewind and reshoot the film of the history of life, the result might be completely different, and our world might be inhabited by unfamiliar beings. The dramatic disappearance of the dinosaurs was not programmed, but the effect was of great significance: the vacated space enabled mammals to evolve, paving the way for the advent of human beings.

2. Natural selection is contingent upon the environmental conditions in which a particular species evolves. One may concede that by favouring those individuals best adapted to the environment in which they live, selection plays a certain, albeit fortuitous role. However, these pro-

cesses cannot be scientifically deduced, given the knowledge and means available today. Their unpredictability is not due to the stochastic nature of the phenomena but to the complexity of interactions between genes and environment about which too little is known.

It is more difficult to invoke pure chance in cases where species have flourished as a result of adaptive radiation. There is seems to be some kind of evolutionary determinism that leads to the appearance of sympatric species, each endowed with a particular aptitude for exploiting certain resources of the milieu. It is almost as if evolution favoured diversification to exploit the capacities of the ecosystem in an increasingly sophisticated manner.

3. The study of ecosystem functioning has demonstrated the existence similar functional processes in different milieus, arousing the suspicion that there might be some kind of 'regulation' after all. This shifts the debate to another plane: are the 'regulatory mechanisms' of an ecosystem deterministic or stochastic? In theory, regulation implies control and an identifiable objective. This is the case for cells or organisms which are systems in pursuit of an intrinsic goal, such as survival and reproduction (deterministic systems). However, when applied to biological systems of an hierarchic order above the individual level, the term takes on an ambiguous note. It has never been possible to demonstrate conclusively that the dynamics of an ecosystem are geared to the pursuit an identifiable objective. Ecosystems are not constructed in the same way as organisms: they comprise lots of organisms theoretically interacting on a stochastic rather than a deterministic basis. Today's ecologists tend to accept that there are 'regulatory mechanisms' at the ecosystem level, analogous to those for organisms, but hold that these mechanisms are generally the result of stochastic events.

The debate remains open, given that:

• Certain interactions between components are 'quasi-deterministic' in the sense that they involve strong coevolutionary relationships (plants and pollinators, for example, or prey/predator relationships). For some authors, like Richard Dawkins, the real object of life is to enable the survival of DNA. 'We are survival robot vehicles blindly programmed to preserve the selfish molecules known as genes.' Viewed from this perspective, the diversity of the living world is a manifestation of the inventiveness of DNA in adopting original techniques to maximize its chances of survival.

- Certain species may replace others without modifying a given functional process.

- The interactions among species within a trophic system could be said to contain deterministic elements of control, if only in the form of the relationship between the eaters and those eaten.

- The functioning of an ecosystem depends strictly upon the spatio-temporal availability and dynamics of environmental factors; in particular: water, nutrition and energy resources.

4 The Species Richness of Natural Communities as a Result of Equilibrium/ Non-equilibrium Processes

The biodiversity of a biotope is not fixed: it is a dynamic attribute of the ecosystem. Biodiversity results from both past and present selective pressures exerted by the biotic, physical, chemical and spatial characteristics of an environment. The dynamics of biodiversity cannot be understood without also considering the dynamics of assemblages and ecosystems.

Theories of equilibrium play a large role in most fields of science. The concept of the 'balance of nature' has a long history in scientific literature, and most explanations for patterns of biological diversity are based on some sort of equilibrium, defined as a balance between opposing forces operating on different space and time scales. On the other hand, one of the major ecological insights of the 1970s was the discovery of the importance of non-equilibrium processes in the maintenance of species diversity. Since then, the notion of equilibrium has been much criticized in ecology, with new theoretical debates about the regulation of species diversity. Most of the disagreement over the validity of contrasting hypotheses stems from a failure to define the terms equilibrium and non-equilibrium clearly and to consider the spatial and temporal scales at which specific processes are important. Equilibrium and non-equilibrium dynamics are not mutually exclusive, and how observations are interpreted is usually a question of the scales at which investigations have been done.

Biodiversity   Christian Lévêque and Jean-Claude Mounolou
© 2004 John Wiley & Sons, Ltd   ISBN 0 470 84956 8 (Hbk)   ISBN 0 470 84957 6 (pbk)

4.1   Theories of Equilibrium Based on Interspecific Relationships

Ecologists have worked towards identifying factors that explain the composition and structure of biological communities. In particular, they have tried to establish the role of interactions between species in maintaining population equilibriums within communities. Following the Darwinian train of thought, population ecologists of the 1950s focused on competition between species. Only much later did theories arise that stressed the role of mutualism and cooperative relations between species, without completely rejecting the idea of competition. The theory of insular biogeography proposed by MacArthur and Wilson in 1967 has revived ecological debate by demonstrating the importance of history and chance in the composition of animal and plant communities.

4.1.1   Theories of equilibrium as a result of interspecific competition

The term 'interspecific competition' describes the competition between individuals belonging to different species over the use of nutritional or territorial resources available in limited quantities. As with intraspecific competition, this may cause a decline in the growth, survival or fertility of the particular species involved.

The principle of *competitive exclusion* derives from Darwin's *Origin of Species*. One of the basic axioms of modern ecology, it formulates the intuitive belief that two species with overly similar ecological characteristics cannot coexist over long periods of time. In other words, in a stable environment, whenever two or more species are in competition over the same limited resources, the species that is best adapted will eliminate the other(s).

Hutchinson (1957) basically upheld the idea that competition is the main factor limiting the diversity of species and leads to the emergence of 'patterns' in the structure of communities. He also popularized the notion of 'ecological niches,' which he defined as hypervolumes with n dimensions, each corresponding to a biological or ecological requirement of the species under study. Competition between species may take place over one or more dimensions of the ecological niche. In theory, the more specialized species are, the less their niches will overlap.

Ecological niches

For MacFayden (1957), the ecological niche was 'that set of ecological conditions under which a species can exploit a source of energy effectively enough to be able to reproduce and colonize further such sets of conditions.' Actually, the term has been used as a generalization of the notion of habitat, which is illustrated as a multidimensional hypervolume of resources axes. Later, Odum (1975) defined the ecological niche of a species as the role that the organism plays in the ecosystem: 'the habitat is the "address" so to speak, and the niche is the "profession" '. In other words, the niche of a species corresponds not only to its place in the trophic network but also to its role in the recycling of nutrients, its effect on the biophysical environment, etc. Today, there is a tendency to characterize ecological niches with respect to three main axes grouping most of the variables pertinent to the living environment: the habitat axis (climatic, physical and chemical variables); the trophic axis; the temporal axis (use of food resources and occupation of space over time).

A great many laboratory experiments have been devoted to the model of competitive exclusion; however, it is not very practicable in the field. The methodological difficulties are such that we do not know what roles such interspecific competitive phenomena actually play with respect to other factors in the general dynamics of ecosystems, particularly in ecosystems where disturbances create strong spatiotemporal dynamics. In reality, environments rarely remains constant over long periods, and modifications in ecological conditions modulate the competitive relations among species. Exclusion and coexistence are not straightforward alternatives. Their relationship appears to be subtler. It is more helpful to consider them in terms of trade-offs and evaluate the degree of coexistence possible between species. A dominated species is rarely eliminated entirely, and many observations indicate that the extinction of a species can be a very long process extending over numerous generations.

4.1.2   The role of predation

One of the ways to limit the growth of a population is to control its development through predation. In a certain sense, the role of predation in the regulation of communities is part of the general theory of competition. However, the dynamics of predator–prey systems are complex and depend upon local conditions – both biotic and abiotic.

A well-known model of prey–predator interaction is the Lotka – Volterra model, which simulates the dynamics of abundance for a single prey species and a single predator species under theoretical conditions where the predator has only one prey and the prey population is not limited by the availability of food. Applied to the North American lynx and the snowshoe hare and based on the number of skins listed in the registers of the Hudson's Bay Company over almost one century, the model provides an account of the alternating fluctuations in abundance of predator and prey observed in nature: when the predator population increases, the prey population decreases, and vice versa. This mathematical model predicts that the outcome of prolonged competition between species in a confined space always ends with the elimination of one species and the total victory of the other. In reality, the situation is rarely so simple. A predator may consume different varieties of prey, depending upon their relative availability, without necessarily having a limiting effect on every one of them.

One hypothesis claims that predators only eat the required number of prey without depleting their 'capital' and thus always have enough food at their disposal. This is what Slobodkin (1968) calls the *optimal strategy of predation*. But such a situation cannot be generalized, because there are a number of cases where predators kill beyond the threshold suggested by the *optimal strategy*.

Another theory suggests that by limiting the abundance of prey, predators enable the coexistence of a greater number of species. In a famous experiment on the Pacific coast of the USA, Paine (1966) demonstrated that the experimental removal of a superpredator, the starfish (*Pisaster ochraceus*), enabled mussels (*Mytilus californianus*) to spread over the rocky sea-beds and led to a simplification of the biological system. Under normal circumstances, the mussel population is controlled by the starfish. Originally composed of 15 species, sea-bed communities were reduced to eight species by the end of the experiment, because the mussels eliminated several other species competing for the same space.

4.1.3 Mutualism or co-operative relationships between species

From the 1950s to 1970s, ecological research was largely driven by the idea that competition is the main factor of interaction between species. Since then, the results of many studies have demonstrated that mutualist inter-actions are far more frequent than previously believed. The concept of mutualism had already been introduced by a Russian, Kropotkin, in a work entitled *Mutual Aid*, first published in 1906. Kropotkin criticizes Darwin for overemphasizing the 'struggle for life', in line with the ideas of Malthus, and overlooking the fact that in nature, living things co-operate towards ensuring their collective survival. They appropriate territories collectively, practising mutualism rather than competition.

Without going into detail, several examples of mutualism are as follows:

- the relationships between plants and micro-organisms, e.g. mycor-rhizeae associated with roots, that facilitate the transfer of nutrients to vascular plants;

- the role of insects in the pollination and dispersal of seeds;

- the roles of mammals and birds in disseminating fruits and seeds in tropical forests (zoochory);

- the lichen formed by the association of one alga with one fungus;

- symbiotic protozoans, able to hydrolyse cellulose, living with social insects (ants) or in the stomachs of ruminants;

- the association between corals and zooxanthellae (unicellular algae).

4.1.4 Saturation of communities and biotic interactions

If interspecific competition and other types of interactions really act as structuring forces within biological communities, it should be possible to find situations in which such interactions limit the number of species that can coexist locally in a stable community. In species-rich biogeographic zones, for example, the specific richness of communities could be expected to reach a certain limit or point of saturation on the local level, regardless of how many species are present in the region. In theory, two kinds of actual results are anticipated (Figure 4.1).

- There is a linear relationship between local and regional richness. In other words, local specific richness is independent of biotic inter-actions occurring at the local level and increases proportionately with the specific richness of the region. This situation is one of

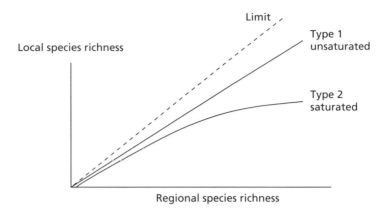

Figure 4.1   Theoretical relationship between local specific richness and regional specific richness in saturated communities (asymptotic curve) and unsaturated communities (linear relationship)

unsaturated communities in which the interactions between species are not important enough to limit specific richness at the local level.

● There is an upper limit to specific richness at the local level. The number of species may at first increase with the increase in regional specific richness, but will eventually stabilise and become independent of the regional specific richness at a certain level. In this case, the curve is asymptotic, and we have what is called a saturated community, where it becomes difficult to add species without eliminating others.

Since it is difficult to test these hypotheses under natural conditions, ecologists use indirect methods. A relatively simple test to verify the hypothesis of saturation consists in comparing local specific richness with regional specific richness in similar habitats belonging to different geographic zones and inhabited by different pools of species. The hypothesis of non-saturation has been tested on fish communities in west African rivers belonging to the same biogeographic zone (Figure 4.2). There is a clear relationship between the surface area of the river catchments and their specific richness, varying between 18 and 95 species. The analysis of local richness of fish species observed in different tributaries of these rivers reveals a close relationship between local specific richness and the specific richness of the river, thus demonstrating that the communities are not saturated. In other words, the factors controlling specific richness on the

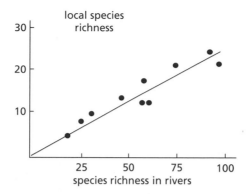

Figure 4.2 Relationship between specific richness in fish in tributary streams and the number of fish species present in the bodies of west African rivers (based on Hugueny & Paugy, 1995)

scale of the entire river are the same as those controlling the structure of the communities at the local level.

4.1.5 The MacArthur–Wilson Model and the theory of dynamic equilibrium

The theory of dynamic equilibrium developed by MacArthur and Wilson (1963, 1967) was first applied to insular environments. The basic hypothesis is simple: the species richness of insular communities depends upon the balance between the rate of immigration and the rate of extinction, expressed as number of species per unit of time. The immigration rate for new species decreases as the number of species settled on the island increases. On the other hand, as the number of species present increases, competitive interactions on the island tend to increase the number of extinctions. The interaction between these two dynamic processes provides an explanation for the specific richness of insular communities (Figure 4.3). One speaks of dynamic equilibrium, because the community is not fixed *a priori*, and species may be replaced over time. More precisely, the MacArthur–Wilson Model predicts:

- that islands close to continental sources will have more species than islands farther out; moreover, certain species have a greater capacity for colonization than others;

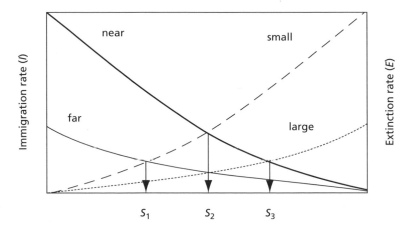

Figure 4.3 Model of the dynamic equilibrium of specific richness on an island. The rates of immigration (I) and the rates of extinction (E) depend upon the distance from the mainland (near or far) and the surface of the island (large or small). The specific richness at the point of equilibrium (S) is given by the intersection of curves I and E (based on MacArthur and Wilson, 1967)

- that species richness is higher on large islands than on small islands, given equivalent positions with respect to the source of invasion; the relationship between surface area S and number of species N is expressed by the formula:

  $N = kS^Z$ or $\log N = \log k + z \log S$

The existence of such a relationship between area/number of species is an empirical fact. There is nothing to tell us why this should be so. However, some scientists believe that the low number of species on small islands is the result of a higher rate of extinction due to the small population size. But an equally plausible explanation is that the diversity of habitats increases with the catchment area, and that communities are richer where habitats are more diversified.

This theory on insular biogeography has a number of weak points. For one, it ignores the fact that not all species have the same biological characteristics and that therefore, they do not have the same opportunities for colonization, at least not in the short term. In the original theory, all species were considered identical, without differentiating how individual species contribute towards balancing immigration and extinction rates on any given island. Moreover, the theory can only be strictly applied to species that can move easily from one place to another, whether on their own (birds, insects, etc.) or else borne by the wind or other species. It should also be added that the theory leaves no room for *in situ* speciation and ignores relationships between species and characteristics of the environment, nor does it take into account the size of the populations present. In spite of all these reservations, this simplified theory has considerable heuristic value and has had significant impact upon research on the organization and dynamics of communities. It has also been applied to continental situations: a mountain summit, a forest grove, a river or lacustrine basin can be treated as 'continental islands.'

Continental islands

An *in situ* experiment, carried out by Wilson and Simberlof (1969), corroborated the theory of insular biogeography to a certain extent. After compiling an exhaustive inventory of the arthropods inhabiting the small islets of mangroves along the coasts of Florida, the fauna of these islets was exterminated. Recolonization occurred rapidly, and after 200 days, the number of species stabilized at values close to those registered at the onset of the experiment. However, the taxonomic composition of the new communities differed from the original, indicating that there is an element of chance in colonizations, with early colonizers enjoying an advantage over latecomers. Despite these taxonomic changes, the trophic structures of the new communities were similar to those of the original communities, with the various trophic groups (herbivores, decomposers, predators) represented in the same proportions as before. This is a point in favour of the hypothesis that there may be redundant species fulfilling similar functions within a community.

4.2 Theories of Succession

An ecosystem is not a fixed structure, established once and for all time. As with living beings (but the analogy ends there), it is a structure that is born, that develops, acquiring complexity and properties, and eventually dies and disappears. But, in contrast to living beings, an ecosystem can also be rejuvenated or change its individual characteristics.

4.2.1   Succession

The emergence of the concept of succession at the beginning of the 20th century introduced the dimension of time into what had hitherto been a somewhat static perception of ecology, paving the way for research on the temporal dynamics of communities. The term succession is used to designate the process of colonization of a biotope by living organisms and the changes in floral and faunal composition that gradually take place in a biotope following a disturbance that has destroyed part or all of the pre-existent ecosystem.

Succession may theoretically proceed by the following stages.

- In a newly created, virgin environment (*juvenile ecosystem*) or an environment that has just experienced a disturbance eliminating most of its species, so-called pioneer or opportunistic species will be the first ones to develop. These species are characterized by high fecundity and rapid population growth (demographic strategies of type 'r') and are not very specialized. Their trophic networks are simple.

- The biocenosis becomes more diversified with the apparition of species characterized by slower growth rates (demographic strategies of type 'K'); food chains become more complex.

- At the *mature stage*, species richness reaches a maximum, including lots of slowly growing species with high life expectancies. The web of interactions and trophic network are complex. Productivity is low, and a large proportion of matter is recycled on location.

- Ecosystems can sometimes be described as *ageing*, particularly when a small number of species get the upper hand over the others and eliminate them. The process whereby a lake gradually deteriorates and finally disappears, giving way to plant formations, is one particular manifestation of this aging process.

A fundamental characteristic of succession is its reversibility. A disturbance may result in the disappearance of all or part of species from the site. If the disturbance is severe and occurs at a mature stage, the cycle of succession is reinitiated, and the ecosystem is 'rejuvenated.' This process may occur repeatedly. Disturbances may take place either as cyclical or as chance

Demographic strategies types 'r' and 'K'

Natural populations have developed numerous morphological and physiological adaptations to the environments in which they live. These adaptive strategies are linked to demographic characteristics: fertility rate; mortality rate, reproductive maturity, life expectancy at birth, etc. What is at stake is, quite simply, the survival of the species. Two extreme strategies for survival have been identified, with most species situated within the spectrum in between: 'r' strategies and 'K' strategies. These terms refer to the two parameters of the S-shaped logistic growth curve of populations: the growth coefficient 'r' describes initial exponential growth of a population, whilst 'K' represents the maximum value of its biomass, a value that it will tend to approach asymptotically.

In schematic terms, one might describe these strategies as follows:

- **r-strategy**: small-sized species; a short life cycle with fast multiplication and thus a rapid turnover rate. These are expanding populations with early reproductive maturity that are generally able to utilize resources (food, space, etc.) rapidly and are limited only by the availability of resources. They are in competition with other species over resources. These mobile and wandering species colonize variable and unpredictable environments, i.e. 'young' ecosystems.

- **K-strategy**: the demographic characteristics of 'K' populations are in contrast with 'r' populations: species with long life cycles and low turnover rates. These are sedentary populations that reproduce late and occupy specialized niches. The species co-occurring in a biotope do not use the same resources, so they are not in competition with one another. They have a high life expectancy and low juvenile mortality rate, often involving protection of the young. These species use energy more economically than pioneer species. They settle in the wake of pioneer species and tend to represent more advanced stages of succession.

In schematic terms, r-strategists are opportunistic species, colonizing instable environments, while K-strategists are 'climax' species.

events, with varying intensity. In other words, an ecosystem may pass through several such phases of rejuvenation, with the biocenosis returning each time to a stage that more or less resembles its pioneer stage. If the frequency of disturbances is too high, the ecosystem may remain in a juvenile stage permanently. The dynamics of such successions are a function of the time intervals between disturbances relative to the life cycle durations of the species constituting the biocenosis. Species with rapid life cycles (r-strategy) recolonize quickly, while species with slow life cycles (K-strategy) will find no opportunity for recolonization if the interval between two perturbations is shorter than the duration of their life cycle.

4.2.2   The concept of climax

In the early 20th century, Clements, an American ecologist placed the concept of climax at the centre of ecological theory (Clements, 1916). In its original sense, *the climax is the ultimate steady-state achieved in the evolution of the vegetation of an ecosystem*, following a succession of intermediary stages, and in the absence of natural or man-made disturbances. The climax community for any given region was thought to be determined by climate and soil conditions. The point is not to classify groups of plants as in phytosociology, but rather to understand the ecological factors driving the evolution of terrestrial vegetation towards a steady-state in accordance with the regional climate. In temperate climate zones of western Europe, for example, the climactic climax corresponds to a mixture of different types of central European and Atlantic oak trees, mountain beeches, and sub-Alpine coniferous forests (spruces, larches, mountain pines).

The concept of climax, as applied to ecosystems, reflects the quest for equilibrium that motivated population ecologists for a long time. However, this initial view was later challenged by other ecologists, who interpreted communities as random assemblages of species that would inevitably change in the long run.

4.2.3   The Holling Model

Since the 1980s, ecologists have been questioning the idea that biological communities pass through a well-ordered, one-directional sequence of stages in their evolution towards a climax whose characteristics are determined by climatic and edaphic conditions. Research on different types of ecosystems has revealed that:

- there is a significant degree of chance involved as to which species colonize an ecosystem after a disturbance or in the course of succession;

- pioneer species and species that characterize mature stages can both remain present throughout the course of succession;

- disturbances like fire, wind and herbivores are major drivers of the inner dynamics of systems and often lie at the origin of succession cycles;

- different disturbances may propel ecosystems towards different ranges of stability; i.e. there is more than one possible climax.

The notion of climax is altogether useful, but it yields an overly static and incomplete view of the phenomenon. Holling (1986) therefore proposed a model that identifies four main periods in the cycle of terrestrial ecosystems (Figure 4.4). This view of succession turns upon an adaptive cycle consisting of two long periods of slow accumulation and transformation of resources (a phase of *exploitation*, corresponding to the rapid colonization of recently disturbed areas, and a phase of *conservation*, with the gradual accumulation and storage of energy and matter), alternating with two shorter periods that create opportunities for innovation (a phase of *release* of the accumulated biomass and nutrients by fires, epidemics, etc., followed by a phase of *reorganization* of the ecosystem). The phase of conservation, which corresponds to a period during which the system becomes more stable and accumulates biomass, may be likened to the concept of climax, but it is embedded in a more dynamic view of ecosystems.

4.3 The Dynamic Equilibrium of Ecosystems and the Role of Disturbances

For a long time, for both practical and conceptual reasons, ecologists studied homogeneous ecosystems that were more or less independent of one another. Reality tells a different story; natural environments are heterogeneous, often fragmented, and they change over time. Ecologists

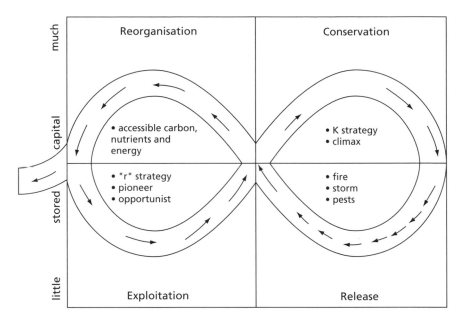

Figure 4.4 Ecosystem dynamics indicating the transitions among the four stages of an ecosystem. The arrow shows the speed of that flow in the cycle, with arrows close to each other indicating a rapidly changing situation and arrows far from each other indicating a slowly changing situation. The cycle reflects the changes in the amount of accumulated capital (nutrients, carbon, etc.) stored in variables that are the dominant structuring variables at that moment in the system. (based on Holling, 1986)

are now agreed that variability and heterogeneity are among the fundamental characteristics of ecosystems and guarantee the long-term stability of communities.

Long-term studies of ecosystems show that their state at any given time depends upon a combination of their history and the current dynamics of their environment. Over longer periods of time, they tend to oscillate around a median state with varying regularity and amplitude. Rather than remain in a so-called steady-state, an ecosystem is actually an interactive system: a change in the environment triggers a dynamic response throughout the whole system, including numerous positive or negative feedbacks. On the other hand, the ecosystem may also have influence upon its environment, implying a reciprocal relationship; the former is not entirely subordinate to the latter.

4.3.1 What is a disturbance?

The important role that disturbances play in the dynamics of communities and ecosystems is one of the most interesting factors to shed new light on ecological paradigms in the last twenty years. A general definition of disturbance was advanced by Pickett and White (1985) and modified by Resh *et al.* (1988): 'any relatively discrete event in time that is characterized by a frequency, intensity and severity outside a predictable range, and that disrupts ecosystems, community or population structure, and changes resources, availability of substratum, or the physical environment'. For Townsend (1989), a perturbation is 'any relatively discrete event in time that removes organisms and opens up space which can be colonised by individuals of the same or different species.' As for Sousa (1984), he holds a disturbance to be 'a discrete, punctuated killing, displacement, or damaging of one or more individuals (or colonies) that directly or indirectly creates an opportunity for new individuals (or colonies) to become established'.

Predation is not, strictly speaking, a form of disturbance, because it is intrinsic to the life of the prey species that must adapt to it. Nevertheless, voluntary introduction of new predator species into an ecosystem can be considered a perturbation of that ecosystem, because new predators and their prey have obviously not had the time to coevolve.

Disturbances can be qualified by different descriptive characteristics: type (physical, biological, etc.), pattern of occurrence (spatial distribution, frequency, intensity, duration, etc.) and regional context. Depending upon their nature and intensity, certain disturbances elicit no response from the ecosystem, while others, for example destroying a habitat and its settlement, may qualify as catastrophes. More generally, a disturbance usually leads to a general restructuring of the ecosystem. In this sense, it can be seen as a rejuvenating process within the phenomenon of ecological succession.

4.3.2 The intermediate disturbance hypothesis

Hutchinson began as an ardent supporter of theories of competition, but several years later, he observed that the model did not apply to phytoplankton communities in lakes. In a famous paper (Hutchinson, 1961), he describes what he calls the plankton paradox: many more species of phytoplankton have been observed to coexist in a relatively simple environment than are accounted for by the theory of competition and the limitation of resources. He therefore reverses the logic of the question: could the great species richness observed not be the consequence of fluctuations in the environment that prevent it from attaining a state of equilibrium over time? If this were so, then coexistence would be the result of non-equilibrium phenomena, rather than characteristic of a state of equilibrium. In a context where equilibrium theories are the predominant paradigm, such novel ideas have been slow to receive the attention they deserve. But they point the way towards future discoveries about the impact of perturbations upon the specific richness of ecosystems.

Already, many ecological investigations have reached the conclusion that disturbances might actually work in favour of biological diversity by lowering the pressure of dominant species upon other species and allowing the latter to develop. Proposed by Cornell (1978) 'the intermediate disturbance hypothesis predicts that species richness will be greater in communities with moderate levels of perturbation than in communities without any disturbances whatsoever or communities that are subject to overly large and/or frequent disturbances'. The initial objective of this hypothesis was to explain the high specific richness of tropical forests and coral reefs. Where disturbances are infrequent, interspecific competition limits the number of species likely to become established, and the most competitive species (resident species) occupy the available space. Con-

---

Uprooted trees: a factor in the maintenance of forest biodiversity

In forest milieus, natural regeneration frequently succeeds the disturbance created by the accidental uprooting of a tree: a hollow forms in the place where it formerly stood and the local exposure of undergrowth to sunlight is increased. The fall may be the result of violent winds or dying tree roots. Such events occur in a chance distribution over space and time. They favour the establishment of 'new' species and contribute to the maintenance of biological diversity. The recolonization of these spaces is accomplished (i) by the assemblage of seeds that arrived at the spot before the disturbance occurred and were dormant in the soil, pending the arrival of favourable light conditions; (ii) by seeds coming from external sources after formation of the hollow; and (iii) by offshoots of the trees.

versely, when perturbations are frequent and/or intense, the dominant competitive species are eliminated, and only colonizing species with brief life cycles are successfully able to maintain themselves. If the disturbances occur with moderate frequency, intensity and amplitude, resident species will cohabit with pioneer species, resulting in the greatest specific richness.

4.3.3 Buffering and recuperative capacities of ecosystems

The buffering capacity of an ecosystem resides in its ability to activate internal mechanisms that dampen the impact of a disturbance. But what about ecosystems' capacity to recuperate after disturbances? Although there has been great emphasis on evaluating the impacts of different disturbances upon the structure or functioning of ecosystems, little research has been done on long-term processes of recovery of disrupted ecosystems.

The capacity of ecosystems to recover from severe disturbances depends upon the biological characteristics of their constitutive populations: rate of generation, fecundity, capacity for dispersal, etc. (cf. box on Demographic Strategies). But there are other factors that are independent of the nature of the organisms involved, such as modifications in habitats, residual toxicity, the duration and period of the impact in relation to the reproductive cycle, the existence and distance of refuge zones, etc. Certainly, an ecosystem's recovery capacity also depends upon the existence of refuge zones for flora and fauna, providing the reservoirs for recolonization after the disturbance.

Oil slicks

Oil slicks are spectacular events punctuating the history of crude oil transportation by sea. What can we learn from these involuntary field experiences, which are seized upon by the media and depicted as ecological catastrophes of huge proportions?

The study of the medium and long-term consequences of hydrocarbon pollution illustrates the importance of the phenomena of ecological succession and regulation for the restoration of disturbed ecosystems (Laubier, 1991). In extremely simplified terms, once the phase of acute toxicity has passed, the consequences of oil slicks are

continues overleaf

Oil slicks (*continued*)

comparable to those elicited by organic waste. Following the accident involving the *Amoco-Cadiz* (March 1978), areas directly exposed to pollution suffered immediate and total destruction of their macrofauna. After several months, when the hydrocarbons had lost their toxicity, the areas were resettled by replacement fauna adapted to anaerobic conditions and to sediments rich in organic matter. These opportunistic species began to proliferate, including some species characterized by large biomass and high production. Gradually, other tolerant species took hold, followed by more sensitive forms. After 4 years, the communities were restored, and in the following years, the group of sensitive species became predominant. After 7 years, the composition and structure of the communities closely resembled the situation existing before the disturbance. Of course, the recuperative ability of marine systems in no way justifies the occurrence of such events.

4.4   Spatial Heterogeneity and Temporal Variability

Spatial heterogeneity and temporal variability are major characteristics of ecological systems. However, for a long time, ecologists underestimated their importance. Until quite recently, they did not have the conceptual and methodological tools to integrate heterogeneity into their research in any practicable way.

4.4.1   Fragmented communities

In an environment characterized by spatial heterogeneity and/or fragmentation of ecosystems, populations of the same species tend to be fragmented and more or less isolated from one another. Since the 1980s, the theory of insular biogeography has prepared the way for the theory of metapopulations, which has been the basis for many theoretical and empirical studies on the effects of habitat fragmentation upon populations.

A *metapopulation* is a group of geographically more or less isolated subpopulations interconnected by individual exchanges that contribute toward maintaining the gene flow between the different subpopulations. A relatively simple instance is that of populations inhabiting oceanic or continental islands with continuous or occasional exchanges among them. In a metapopulation, certain subpopulations where births outnumber deaths act as 'sources' from which individuals disperse towards other areas. Conversely, certain subpopulations live in harsh environments where mortalities exceed births. Such environments constitute 'sinks.' A metapopulation is thus a dynamic system characterized by migration flows and processes of extinction between and within the subgroups of a fragmented habitat. The concept of metapopulation can be extended to multispecific communities, a metacommunity being defined as a group of ecological units sharing certain biotic components and among which exchanges are possible.

4.4.2 The dynamics of non-equilibrium

Proceeding from the intermediate disturbance hypothesis, Huston (1979) proposed a model of 'dynamic equilibrium' (Figure 4.5). It is applicable

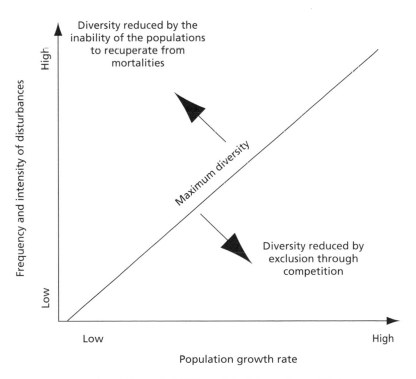

Figure 4.5 Huston's (1979) model of dynamic equilibrium

to species belonging to the same functional group and in competition with one another within a given habitat. The model is based on the relative roles of two opposing forces that both influence species diversity: (i) competition leading to exclusion in situations characterized by high population growth rates and a mild degree of disturbances; (ii) severe or frequent perturbations, causing mortalities and limiting the reproduction of populations with slow growth rates, to the point where such species have no time to re-establish themselves. In the real world, many systems are disrupted frequently enough (fires, tornados, droughts, etc.) to put the recovery of their communities at recurrent risk.

The model does not predict a constant level of biological diversity for a certain combination of parameters, but rather a pattern of fluctuations within limits established by the model of dynamic equilibrium. This model is analogous to that of the theory of insular biogeography (where the opposing forces are immigration and extinction) but is applicable on a smaller spatial and temporal scale.

To date, the dynamics of non-equilibrium provide the best explanation for the spatial and temporal heterogeneity observed in most ecological systems. Disturbances and heterogeneity are the interdependent factors that create opportunities for recolonization and control the structure of communities. Within the context of a perpetually changing environment, competition, predation and interspecific relationships generally appear to have little influence upon the structure of communities, except under certain circumstances and in limited periods of relative stability.

4.4.3   Patch dynamics

Landscape ecology provides ecologists with a simplified view of heterogeneity by defining space as a mosaic of 'patches' (spatially delimited structures at a given time) arranged over an ecologically neutral matrix. This spatial model grows progressively more complex when the varying characteristics of different patches and their temporal dynamics are taken into account.

The *patch dynamics concept* provides the link between the mosaic distribution of communities (metacommunities) and the spatial/temporal dynamics of the patches. Patches may either disappear or grow over time, as a function of fluctuations in environmental factors. Depending upon the prevailing tendency, associated communities can be either senescent, or pioneers or correspond to a stage in a succession cycle. Moreover, each patch of the matrix and its communities may have completely different dynamics from the others. The seasonal inundations and flood recession of river beds are a good illustration of patch dynamics in that variations in water level create and/or modify the spatial heterogeneity of the river channel.

According to Townsend (1989), the concept of patch dynamics is a major unifying principle in the ecology of running waters, where ecological characteristics such as current speed, substrates and availability of resources tend to manifest considerable spatial heterogeneity. Patch dynamics imply the following principles.

- Natural or man-made disturbances act upon ecosystems to modify the distribution of habitats over time and space. To give a simple example, river floods create new aquatic habitats as well as modify ecological conditions in already flooded habitats: current speed, depth, etc. The reverse applies equally when the waters recede.

- In a heterogeneous system, pioneer populations take hold as soon as habitats become available and usually evolve towards a more mature state. Thus, different patches will find themselves at different stages of ecological succession at the same time, as a function of the chronology of inundations.

- Because they are dynamic over space and time, the assemblages are able to maintain a much greater biological diversity than systems evolving towards a climax in a monotonous way. Such dynamics may allow the co-existence of several species, with different ecological needs, in different patches, that are at different stages of evolution.

Thus, spatial heterogeneity and temporal variability are actually key elements in ecosystem functioning and the structuring of communities, and not just a simple 'background noise' that disturbs population dynamics.

4.4.4 From the continental to the local level: a conceptual model to explain the richness of ichthyic communities

Phenomena occurring on large spatial and temporal scales provide a partial explanation for the composition of local communities. The example of river basins (the continental equivalents of islands in the sea) provides a good illustration for scales of interaction and the long-term consequences of certain events. The qualitative and quantitative composition of the fish communities inhabiting river basins is actually the result of numerous past events interacting with contemporary ecological factors. Tonn (1990) proposed a theoretical framework based on the principle that the local composition of species is the result of a series of filters acting on different scales of time and space. Species must have passed successfully through these successive filters to be present in the basin under consideration (Figure 4.6).

Historic events are sometimes expressed on a global scale. For example, the continental filter corresponds to the separation of Gondwana, since the present fish fauna on each of the continents are the result of processes of speciation and extinction that occurred subsequently.

On a regional scale, more recent events either gave or denied species the possibility to spread and colonize the river catchment, or else, they

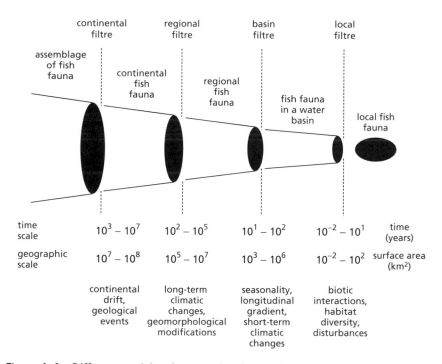

Figure 4.6 Different spatial and temporal scales on which the species composition of fish populations in watersheds can be interpreted (based on Tonn, 1990)

may equally have driven the species once present in these basins to extinction. The different biogeographic zones recognized today can be explained by a succession of many climatic events, occurring over millions of years, and the existence of refuge zones where certain species have been able to survive. In each region, river catchments are presently isolated from one another by terrestrial barriers that fish are unable to cross. However, in the more or less remote past, the basins did communicate with one another through river catchments, orogenetic events or major floods. The current species richness of the river basins is a function of the diversity of their habitats, and biotic interactions play a much more important role on this scale than they do on others.

The number of fish species in a watershed is thus the result of an equilibrium among: processes of colonization and extinction, depending in part upon the past history (climatic, geological, etc.) of the basin; processes of speciation, resulting from the evolutionary potential of the families present and the duration of their isolation; and competitive phenomena and/or epidemics. The model proposed by Tonn is actually an extension of the concept of patch dynamics on the global level and over very long periods of time.

The effect of continental richness upon the specific richness of rivers has been demonstrated. For example, given comparable environmental characteristics (surface area, discharge, regional productivity), South American rivers are richer than African rivers. This is probably the result of historical factors that favoured a greater diversification of species in South America after this continent separated from Africa.

4.5 Are Ecological Communities Governed by Niche-assembly or Dispersal-assembly Rules?

There are two conflicting hypotheses on the nature of ecological communities (Hubbell, 2001). One of them is the view that communities are groups of interacting species whose presence or absence and even relative abundance can be deduced from 'assembly rules', based on the ecological niches or functional roles of each species. According to this view, the stability of the community and its resistance to disturbance derive from the adaptive equilibrium of member species, each of which has evolved to become the best competitor in its own ecological niche. Niche-assembled communities are limited-membership assemblages in which interspecific competition for limited resources and other biotic interactions determine which species are present or absent from the community.

The theory of island biogeography asserts that island communities are dispersal-assembled, not niche-assembled. If communities are largely opportunistic collections of species whose biogeographic ranges happen to overlap for historical and individualistic reasons, then it follows that species in communities are not highly coadapted or codependent. This view does not deny the obvious existence of niche differentiation. However, it breaks away from the conventional neo-Darwinian view of ecological communities as coadapted assemblages of niche-differentiated species residing at or near adaptive and demographic equilibrium. This is replaced by a new world view in which ecological communities are seen to be in perpetual taxonomic nonequilibrium, undergoing continual endogenous change and species turnover through repeated immigrations and local extinctions. It also ascribes much less importance to niches in regulating the relative abundance and diversity of species in the community. MacArthur and Wilson's theory raises the possibility that history and chance could play an equal if not larger role in structuring ecological communities than do niche-based assembly rules.

Actual ecological communities are undoubtedly governed by both niche-assembly and dispersal-assembly rules, but the important question is: what is their relative quantitative importance? Applied ecology and conservation biology and policy critically depends on which perspective is closer to the truth, a fact that is not as widely appreciated as it should be.

5 Biological Diversity and the Functioning of Ecological Systems

In simplified terms, ecologists study three major types of processes involving the living world within ecosystems:

- trophic relationships between groups of organisms (*food chains* or *trophic networks*);

- the role of species in the dynamics of *biogeochemical cycles*;

- *biological production*, i.e. the capacity to produce living matter and thus accumulate energy within an ecosystem.

5.1 Biological Diversity: a Dynamic System

The role of biological diversity in an ecosystem concerns three levels of integration in the living world.

- *Intraspecific diversity*, i.e. the genetic variability of populations. It is due to the genetic diversity which is their biological heritage that species are able to respond to changes in the environment.

- *Diversity among species* in terms of their ecological functions within the ecosystem. Species exist in a large variety of forms, with different sizes and biological characteristics. Operating individually or in groups within trophic webs, these properties influence the nature

Biodiversity   Christian Lévêque and Jean-Claude Mounolou
© 2004 John Wiley & Sons, Ltd   ISBN 0 470 84956 8 (Hbk)   ISBN 0 470 84957 6 (pbk)

and magnitude of the flow of matter and energy within the ecosystem. The different interactions among species, not only competition but also mutualism and symbioses, contribute collectively to the dynamics of an ecosystem.

- *Ecosystem diversity*, corresponding to the variety of habitats and their variability over time. Specific richness is usually considered a function of the diversity of habitats and the number of potentially available ecological niches. Owing to their biological diversity, ecosystems play a global role in the regulation of geochemical cycles (fixation, storage, transfer, recycling of nutrients, etc.) and the water cycle.

In the ecological sense of the term, biological diversity results from dynamic interactions within and among the levels of organisation of the living world, as well as with the physical and chemical environment that it contributes towards modifying (Figure 5.1). The functioning of ecosystems and their flows of matter and energy are thus reciprocally controlled by physical, chemical and biological processes.

BIODIVERSITY

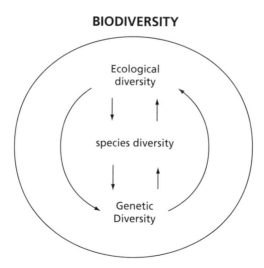

Figure 5.1 The concept of biodiversity involves all the interactions between the diversity of species, their genetic diversity and the diversity of ecological systems (based on di Castri and Younès, 1996)

5.2   The Ecosystem Approach

The Convention on Biological Diversity defines an ecosystem as 'a dynamic complex of plant, animal and micro-organism communities and their non-living environment interacting as a functional unit'.

Its description of the ecosystem approach stresses that an ecosystem can be a functional unit at any spatial scale. It also observes that humans are an integral part of many ecosystems and notes that, because of the frequently unpredictable nature of ecosystem responses and our incomplete understanding of ecosystem functioning, application of the ecosystem approach will require adaptive management techniques. It further states that the ecosystem approach does not preclude other management and conservation approaches, such as protected areas and single-species conservation programmes, but rather that all these approaches could be integrated to deal with complex situations.

At an early stage, the Conference of the Parties decided that the ecosystem approach should be the primary framework of action under the Convention. A document endorsed by the Convention gives a description of the ecosystem approach and a set of twelve guiding principles for its application, together with five points of operational guidance. The Conference of the Parties has asked the Parties to strengthen regional, national and local capacities for the ecosystem approach by identifying case studies, by implementing pilot projects and organising workshops and consultations to enhance awareness and share experiences.

The following points are proposed as operational guides to applying the 12 principles of the ecosystem approach (Table 5.1):

- focus on the functional relationships and processes within ecosystems;

- enhance benefit-sharing;

- use adaptive management practices;

- implement management actions at the scale appropriate for the issue addressed, with decentralisation to the lowest appropriate level;

- ensure cooperation across sectors.

Table 5.1   The 12 guiding principles of the ecosystem approach

Principle 1	The objectives of management of land, water and living resources are a matter of societal choice.
Principle 2	Management should be decentralised to the lowest appropriate level.
Principle 3	Ecosystem managers should consider the effects (actual or potential) of their activities on adjacent and other ecosystems.
Principle 4	Recognizing potential gains from management, there is usually a need to understand and manage the ecosystem in an economic context. Any such ecosystem-management programme should:

- reduce those market distortions that adversely affect biological diversity;
- align incentives to promote biodiversity conservation and sustainable use;
- internalize costs and benefits in the given ecosystem to the extent feasible.

Principle 5	Conservation of ecosystem structure and functioning, in order to maintain ecosystem services, should be a priority target of the ecosystem approach.
Principle 6	Ecosystems must be managed within the limits of their functioning.
Principle 7	The ecosystem approach should be undertaken at the appropriate spatial and temporal scales.
Principle 8	Recognising the varying temporal scales and lag-effects that characterize ecosystem processes, objectives for ecosystem management should be set for the long term.
Principle 9	Management must recognize that change is inevitable.
Principle 10	The ecosystem approach should seek the appropriate balance between, and integration of, conservation and use of biological diversity.
Principle 11	The ecosystem approach should consider all forms of relevant information, including scientific and indigenous and local knowledge, innovations and practices.
Principle 12	The ecosystem approach should involve all relevant sectors of society and scientific disciplines.

5.3   Functions of Species Within Ecosystems

Species differ from one another in the ways in which they use and transform resources, in their impact upon their physical and chemical environment and in their interaction with other species. They are characterized according to their *ecological niches*.

> ## Ecological niches
>
> The American ecologist Odum defined the niche of a species in terms of its role within the ecosystem: 'the habitat is the "address", so to speak, the niche is the "profession".' The niche includes not only the place of the species within the trophic network, but also its role in the recycling of nutrients, its impact upon the biophysical environment, etc.

5.3.1 Keystone species

The concept of keystone species postulates that certain species are more important than others within the network of interactions of an ecosystem. A keystone species could be identified as that species whose loss would cause significant changes in the general structure and processes of the ecosystem.

Key predators are identified as those species whose presence severely limits the presence of other species. Planktivorous fish belong to this category (they limit the abundance of large-sized zooplankton in lakes), as do the large terrestrial predators, whose disappearance in Europe has led to the proliferation of crop pests. *Key mutualists*, on the other hand, are organisms directly or indirectly necessary for the maintenance of other populations. For example, the rate of plant fertilization frequently depends upon the presence of a varied fauna of pollinators (mainly insects).

5.3.2 Engineer organisms

These organisms directly or indirectly control the availability of resources to the other species by causing changes in the physical state of their environment. *Autogenous engineers* modify the environment by the very fact of their own physical structure. This is the case, for example with trees and corals, whose own physical structure creates habitats for other species. *Allogenous engineers* modify the environment by transforming its structure. By cutting down trees to construct dams, beavers modify the hydrology and ecology of rivers. Other examples of allogenous engineers are termites, ants and earthworms, that dig and mix the

---

Biological diversity and the dissemination of tropical plants

In forest ecosystems, one particular function of biological diversity is to ensure that plants are disseminated through a multitude of seed dispersal systems using animals as vectors. *Zoochory* is the most frequent form of dispersal found in tropical rainforests. In French Guyana, 80% of all plant species produce zoochore fruits, involving the intervention of 72 (out of 575) bird species and 36 (out of 157) mammalian species, of which 23 are Chiroptera. The dissemination of each plant species depends on the eating habits of the animal vectors, their routes, resting places, etc. Many plants with large seeds depend upon large mammals such as monkeys, big birds (toucans, Agamis), for their dissemination. The disappearance of these intensively hunted animals may therefore result in a decline in plant diversity.

---

soil, modifying its organic and mineral composition, as well as the cycle of nutrients and water drainage.

5.3.3 Functional groups: complementarity and redundancy

It is not always easy to determine precisely what the relative contribution of each species to ecological processes is. 'Functional groups' is the term used to describe sets of species exerting a comparable effect upon a particular process or responding in a similar manner to changes in their external constraints. This may, for example, consist in the group of species that exploit the same category of food resources, or else the group of species involved in major biogeochemical cycles (nitrogen, carbon, etc.).

An ecosystem function may be provided by one single species or by a limited number of species in one particular system, while there may be a large number of species providing for the same function in another ecosystem. 'Functional redundancy' arises when several species occupying the same spatial niche provide for similar functions, even though their relative importance may vary.

5.4   Hypotheses on the Role of Species in Ecosystem Functioning

What do the different species contribute to the flow of matter and energy within ecosystems and to the mechanisms that maintain and regenerate these ecosystems? A number of studies have shown that ecosystems with different biological communities nevertheless exhibit important similarities in the way they function. In other words, while species may play precise roles within an ecosystem, it is possible for different species to fulfil that same function.

Various hypotheses have been proposed to explain the relationships between the nature and richness of species present in an ecosystem and their role in its functioning (Figure 5.2).

1. According to the *hypothesis of diversity-stability*, the productivity of ecological systems and their capacity to react to disturbances increase

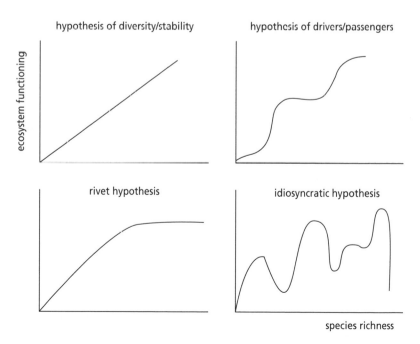

Figure 5.2   Examples of some hypotheses on the relationships between biological diversity and ecosystem functioning

steadily with a rise in the number of species in the system. In other words, the greater the number of interspecific relationships, the better the flow of energy in trophic networks is maintained. If certain relationships are broken as a result of the extinction or disappearance of one or more species, alternative pathways can be established. The corollary is that the ecological functions of different species overlap, such that if one species disappears, its function can be compensated by others.

2. The *'rivet' hypothesis* proceeds by analogy. An airplane wing has more rivets than are actually necessary to hold the wing together. If rivets are progressively eliminated, this may cause the wing to collapse suddenly, once a certain threshold is past. In the same way, an ecosystem's capacity to absorb modifications decreases with the extinction of certain species, even if the performance of the ecosystem appears unchanged. Beyond a certain threshold, a significant change in ecosystem functioning occurs.

    Like the preceding hypothesis, this hypothesis upholds the concept functional redundancy, but at the same time, it emphasizes the existence of specialised functions. In practice, an ecological function does not cease before all the species contributing to that function have been eliminated from the ecosystem.

3. The *hypothesis of 'drivers and passengers'* is an alternative to the preceding hypothesis. It derives from the principle that not all species play equivalent roles. Many species are superfluous (passengers), while only some (drivers) play an essential role in maintaining the ecosystem as a whole. Species with more important ecological functions include, for example, ecological engineers and keystone species. Their presence or absence determines the stability of an ecological function.

4. A final hypothesis (known as the *idiosyncratic hypothesis*) considers the possibility that no relationship exists between the species diversity of an ecosystem and how it functions. Ecosystem functioning is modified when biological diversity changes; however, the magnitude and direction of the modifications are not predictable, because the role of each species is unpredictable and may vary from one area to another.

5.5 Neighbourhood Relationships Between Species

Besides the relationships of eaters to eaten, there exists a whole spectrum of cooperative relationships between species, ranging from competition to mutualism and parasitism.

5.5.1 Competition

Competition involves the struggle of two or more species over the use of the same resource, whether space or food. If the population of one species has a competitive advantage in appropriating a resource, it takes control of that resource and eliminates the populations of other species belonging to the same community. This is the principle of *competitive exclusion*. It may, for example, consist in the competition of plants for light: in a forest, species will grow upwards as quickly as possible in order to capture more light.

5.5.2 Co-operative relationships: commensalism and symbiosis

While ecologists attach great importance to competition between species, they are also interested in cooperative relationships. In fact, prokaryotes rarely function individually in the environment and often maintain mutualist or symbiotic relationships with eukaryote organisms. According to the degree of dependence, the following relationships are distinguished:

- *Commensalism* is a relationship in which one species profits by an association that the other species finds neither advantageous nor inconvenient. This is the case, for example, for epiphytic plants like orchids, for the animals that attach themselves to the shells of molluscs, or the insects that live in rodents' burrows or birds' nests.

- *Mutualism* is a reciprocally beneficial association between two species that may nonetheless lead independent lives. One of the partners plays a role that provides a service to its associate, and because it receives 'compensation' in return, it finds an advantage in the association.

• *Symbiosis* is the term used to designate an association between two species that appears necessary and inseparable. All of the major types of ecosystems existing today have their own cortege of symbiotic relationships, as manifestations of adaptive solutions to different environments. Examples for symbiotic relationships are found equally in the bacteria inhabiting the vertebrate digestive system, and among the coral reefs, where madrepores harbour unicellular algae called zooxanthellae in their tissues. These zooxanthellae live in symbiosis with the polyps, providing them with various organic substances that they produce by photosynthesis.

Lichen provide another good example of a symbiotic relationship. The photosynthetic partner is a form of cyanobacteria or green algae (85 per cent of the time). The other partner is usually a fungus (an ascomycete such as the truffle and morel). While the algae ensure the synthesis of carbon compounds, the fungi provide the algae with a protective substrate and probably facilitate their intake of water and mineral salts. Lichen have colonized almost all terrestrial habitats, including the most hostile, whether they be rocky slopes, mountain summits or arid deserts. They constitute the dominant terrestrial biomass on the Antarctic continent. Altogether, lichen qualify as an evolutionary success story. The key to their survival in extreme climatic environments lies in their ability to survive severe drought stress in a dormant state and quickly re-establish their metabolic activity when they are rehydrated.

5.5.3 Parasitism

Parasitism is a form of relationship in which an organism (the parasite) profits from its host. Parasites sidetrack a portion of the resources normally intended for the growth, survival and reproduction of the hosts and use them for their own benefit. Although they are mostly invisible to the naked eye, parasites are nevertheless omnipresent. They are so very numerous that the question arises whether there may not be more parasitic species than independent species in the world today.

The host–parasite relationship can be depicted as a face-off involving two processes: on the one hand, the probability of an encounter, and on the other hand, the compatibility of the partners. Both the host and the parasite are playing on both sides of the board.

In the host–parasite system, the parasite is constantly trying to invent adaptations that will enable it to encounter its host, whilst the latter responds by elaborating various defences to avoid the parasite or make its life difficult. Thus, both parties are constantly driven to renew their genetic diversity. Parasites achieve this renewal through mutation (like HIV), by gene transfer (as in the case of the influenza virus) and through genetic recombination.

Recent studies have shown that DNA may be exchanged between the genome of the parasite and that of the host, thereby modifying some of the latter's characteristics. This kind of 'manipulation' enables the parasite either to modify the behaviour of the host with a view towards ensuring the parasite's transmission, or to weaken the host's immune system with a view towards increasing the probability of the parasite's survival.

One example of such manipulation is that of freshwater *Gammarus*, small aquatic crustaceans that transmit various parasites to birds when they are ingested. Freshwater *Gammarus* behave differently when they are infested with parasites. Healthy *Gammarus* flee predatory birds by retreating to deep waters when they approach, while *Gammarus* infested with parasites flail about aimlessly at the surface of the water, where they are caught and eaten by the birds. The theory is that this behaviour is the consequence of a neuromediator released by parasites residing near the *Gammarus'* nerve centres.

When a host is invaded by a parasite, its immune system is activated and tries to get rid of the parasite. The host thus becomes a potential killer, and in order to survive, the parasite has to respond by developing alternative strategies of its own: it must either escape the surveillance of the host's immune system, or else develop faster than the immune system can kill. In this way, veritable 'arms races' are waged over generations, amounting to a form of coevolution: each innovation by either partner is reciprocated by an innovation of the other.

Parasites can also be the underlying cause for the selection of life traits in the host: earlier sexual maturity, enabling the host to reproduce before the parasite exerts its detrimental effect; number of offspring; frequency and duration of reproductive periods, etc.

Parasitic or pathogenic organisms often play a role in population control and community structure that is equivalent to the role of preda-

> Numerous species of Hymenoptera have perfected elaborate parasitic strategies with respect to other insects. They deposit their eggs within the body of a caterpillar. The larvae hatch and develop at the expense of the infested host, comparable to a food trough that must be maintained until the larvae are completely developed. In order to obstruct the immune systems of the infested individuals, some Hymenoptera manipulate the physiology of their host. For example, one original strategy involves introducing viral particles into the body of the host, where they cause modifications to growth and development, and obstruct the immune system.

tors. Their impact upon the host's physiological state or viability can be either direct or indirect. For example, there is often a significant drop in the fertility of the infested animals. In the case of the Scottish red grouse, long-term observations have registered periodic collapses in the population due to a parasitic nematode that infiltrates the digestive system. The parasite reduces the fertility of the females and makes them more vulnerable to predators.

Parasitism also intervenes to limit the presence of certain species within a given ecosystem. In the United States, for example, the deer of Virginia are infected by a small nematode, *Elasphostrongylus tennuis*, which is only slightly harmful to them. However, this same parasite is extremely harmful to other ungulate species, such as the moose and the caribou. Since these species suffer high mortalities when they try to invade deer territory in Virginia, the parasite effectively protects the deer from competition with related species. Such situations may be much more frequent than hitherto imagined.

Sustainable interactions

Originating in the early 1980s, the ecology of sustainable interactions, i.e. the study of interactions in the host–parasite environment, arose from the observation that parasites play a major role in the functioning and evolution of the biosphere. In host–parasite systems, two organisms with different genetic information live together, frequently one within the other. The genetic information of each partner expresses itself side-by-side with that of the other in a long-term interaction, as opposed to the short-term interactions typical of predator–prey relationships.

5.6 Food Webs and Trophic Chains

The nature and intensity of the trophic relationships between species living in the same ecosystem play a central role in the circulation of matter and energy. Understanding these relationships is crucial to ecological theory.

In schematic terms, the chain of dependencies whereby some organisms eat other organisms, before being eaten by yet others in their turn, constitutes a *food chain*, or *trophic chain*, and provides a highly simplified description of the circulation of matter or energy through different levels: from autotrophic producers to final consumers. Of course, the reality is far more complex. *Trophic webs* describe the multiple interactions between species, including relationships of eater to eaten, as well as competitive relationships over the use of the same resources (Figure 5.3).

5.6.1 Producers–consumers–decomposers: the flow of matter and energy

In an ecosystem, autotrophic organisms use chemical energy or light energy to synthesise their own organic substance from mineral substances that they draw from the environment. Generally speaking, these *primary producers* include all plants, as well as bacteria and protists. They serve as food for *herbivores*, which are in turn consumed by *carnivores*. All animals, fungi and bacteria that derive their nutrition from living or dead organic matter are known as *heterotrophic* organisms. When organisms die, *decomposers* recycle their organic matter, reducing it to simple mineral elements, that can once again be assimilated by plants.

Prokaryotes manifest greater metabolic diversity than eukaryotes. They include phototrophic species, such as cyanobacteria, that use sunlight as a source of energy, as well as chemotrophs, that draw their energy from chemical substances in their environment. The discovery of bacteria in hydrothermal vents in the deep ocean has shown how bacteria assume the role of primary producers of organic matter in these primal ecosystems, drawing their energy from the oxidation of sulphuric compounds and methane. Various studies have confirmed the importance of chemosynthesis as a generator of life in the oceans.

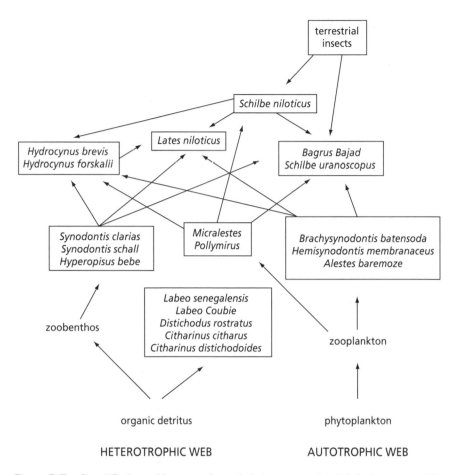

Figure 5.3 Simplified trophic networks and chains supporting ichthyic communities in Lake Chad

The mineral elements essential to the production of living matter are virtually indestructible. In theory, they are perpetually recycled through the ecosystem. The biochemical energy accumulated in the living matter of autotrophs is gradually dispersed to maintain the metabolism and ensure the reproduction of heterotrophic organisms. This is why matter is depicted in a cyclical form, while energy is seen as a flow (see Figure 5.4).

Matter (mineral salts) and energy circulate perpetually from primary producers to herbivores and carnivores, as well as to decomposers. To understand how an ecosystem functions, it is necessary to know what

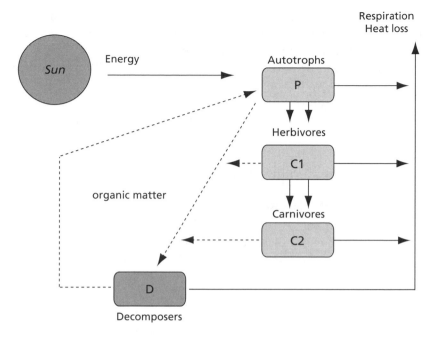

Figure 5.4 Simplified diagram of matter and energy flows in food chains

processes of exchange, transformation and accumulation operate among its living forms.

5.6.2 'Top-down' and 'bottom-up' theories

For a long time, trophic networks used to be considered as linear chains, with the flow of nutritive elements moving up from the primary producers towards higher trophic levels. In this context, it is logical to expect the competition among primary producers for the use of nutritive elements to play a major role in population regulation. This is the theory of community control through resources (*'bottom-up'* *control*).

But many ecologists have demonstrated that there is also an inverse effect and that the functioning of an ecosystem is highly dependent upon the predation practised by higher trophic levels on lower levels (*'top-down' control*). In an aquatic system, for example, fish predation on invertebrates has a considerable impact upon the size structure and specific composition of zooplankton and zoobenthos. When zooplanktivorous fish are abundant, the number of large-sized zooplankton is diminished, benefiting smaller-sized species.

Actually, bottom-up and top-down controls coexist within ecosystems. Their relative importance depends on environmental conditions. Bottom-up control relates more to the level of biological production, while top-down control has more impact upon the structure of communities.

5.6.3   Theory of trophic cascades

The concept of trophic cascades in aquatic environments derives from the principle that an increase in predation by piscivores leads to a decrease in the biomass of zooplanktophagous fish. When the pressure of predation upon zooplankton is reduced, this results in an increase in the biomass of zooplankton, leading, in turn, to a decrease in the biomass of phytoplankton, which is subjected to greater predation.

In theory, if it is possible to adjust the biomass of the predators, it is possible to control the cascade of trophic interactions that regulate algal dynamics. The biomanipulation of aquatic environments consists in modifying algal dynamics by adding or removing predators.

Biomanipulation

The term biomanipulation refers to a deliberate intervention in an ecosystem, undertaken with the aim of restructuring the biological community and achieving a situation that is *a priori* better or more favourable for humans.

The biomanipulation of aquatic systems has been practised in various forms for centuries. Recently, scientists have been concerned with the challenge of modifying specific composition (ecosystem structure) or productivity (functional process) to meet specific objectives, such as, for example, reducing eutrophication or the proliferation of macrophytes, increasing water transparency, fostering a more diversified biological community, etc.

The term covers a whole series of techniques but is most often used to designate a top-down manipulation of fish communities to control algal biomasses though food chains, such as by increasing carnivore species and reducing planktophagous or benthophagous species.

5.7 The Diversity of Species and Biological Production

It is not new to postulate the existence of a relationship between species diversity and ecosystem productivity. However, this fact has not yet been conclusively demonstrated. Environments that are poor in species, such as deserts and tundra, are also systems of low productivity, as compared with species-rich tropical forests. Conversely, humid zones or agricultural systems exhibit high biological productivity coupled with a reduced number of species. Thus, high productivity is not necessarily associated with great biological diversity. In fact, numerous observations appear to demonstrate that the flows of energies in ecosystems are not much affected by the number of species present.

In order to learn more about the role of biological diversity in ecosystem functioning, and given the difficulty of studying this question in the natural world, ecologists have carried out numerous experiments in controlled environments. These experimental set-ups enable researchers to test ecological theories and study natural processes under simplified and controlled conditions. At Ecotron (in the UK), natural environments are simulated in 16 different enclosures, controlling factors such as light, rain, humidity, temperature, etc. These miniature ecosystems may contain up to 30 plant and metazoan species, representing four trophic levels (plants, herbivores, parasitoids and detritivores), interacting over several generations. It is possible to make replicas for statistical analysis.

Experiments have also been carried out in plots in natural situations. The objective of the European BIODEPTH (BIODiversity and Ecological Processes in Terrestrial Herbaceous Ecosystems) project is to confirm or refute the existence of a relationship between specific richness and productivity in grassland ecosystems. The project has eight sites, ranging from Sweden in the north to Greece in the south. Early results indicate a general effect of plant diversity upon biomass production, independent of grassland type and geographic location.

Three general conclusions emerge from different experimental studies investigating the relationship between species richness and ecosystem productivity.

- *Greater specific richness constitutes a form of insurance for long-term ecosystem functioning.* Ecosystems where several species fulfil the same functions (redundant species) appear to be better adapted to respond to disturbances than those in which each species fulfils one, unique function. In other words, if several species exploit the same resources, as is the case for generalist herbivores, the gain or loss of one species will have an effect upon the composition of the communities, but it will have little impact upon the ecosystem processes, insofar as the other species will compensate the change. The behaviour of such communities is quite predictable (cf. the rivet hypothesis, for example);

- However, not all ecologists support the idea that biological diversity causes an ecosystem to function better. According to certain studies, ecosystem responses to such changes depend upon the specific composition of the community and its biological or morphological characteristics. In experiments carried out under controlled conditions, the presence or absence of species more able to use the resources than others (so-called dominant species) has emerged as one important explanatory factor. In reality, it is not so much species richness, as such, that is important, but rather the biological characteristics of the species and the diversity of functional types represented. These qualities are much more difficult to quantify than specific richness. Under these circumstances, it is not so easy to predict how a system will behave in the event of a gain or loss of species (cf. the drivers' and passengers' hypothesis);

- Interactions among species may generate positive or negative feedback at the ecosystem level that combines with previous effects. Given the complexity and variability of the interactions involved, these effects are usually difficult to establish; nevertheless, the importance of such processes should not be disregarded. Particularly in food chains, changes in one functional group may have important consequences for the dynamics and production of other functional groups (see, for example, the theory of trophic cascades).

All these studies point to the conclusion that a greater biological diversity is more favourable to the production and stability of ecosystems and helps to ensure the perpetuation of the cycles of matter and energy. Ecologists have long held this to be true – a conviction that echoes traditional rural beliefs.

How representative are mesocosms?

When ecological experiments are undertaken on a small scale, the question is always to what degree they are representative with respect to the far greater complexity of natural environments. Interactions between biotic and abiotic processes complicate the design and interpretation of ecological experiments. When an experimental manipulation has multiple components, but only one of them is identified as the experimental treatment, conclusions about cause and effect relationships are likely to be erroneous, because the actual cause of an observed response may be disregarded in the interpretation of the results. The unrecognised cause of an observed response can be denoted as a 'hidden treatment' (Huston, 1997). For example: microcosms assemble only a limited number of species without any possibility for exchanges with other milieus; certain processes operate only on larger spatial or temporal scales; some organisms may be too large with respect to the mesocosms. Certainly, experiments in controlled systems, such as those conducted by Ecotron, can never replace observations made in natural environments. Most importantly, they have little pertinence for the study of large-scale processes. Nevertheless, they do provide a way of tackling questions that cannot be dealt with in the natural world. We need a large variety of approaches towards understanding how nature functions; what is important is to remain critical of results obtained.

5.8   Biological Diversity and the 'Stability' of Ecosystems

The term 'stability' is highly contested. It derives from the idea that an ecosystem has a structure and mode of functioning that endure over time, at least on the time scale of human beings. *Persistence* and *permanence* are terms sometimes employed to characterize ecological systems that maintain themselves in this fashion without significant modifications. The term *resilience* (or *homeostasis*) refers to the capacity of an ecosystem to recover its primitive structure after having been subjected to a disturbance.

The question of the relationship between biological diversity and the resilience or stability of ecosystems has been much debated. One more or less intuitive postulate holds that the more diversified an ecosystem is, the more stable it will be. Based upon the existence of redundant species, this hypothesis expresses a simple presupposition: if the number of linkages in an ecosystem increases, then the disappearance of any one linkage will soon be compensated by the development of another.

Some recent results support this hypothesis. Both laboratory and field experiments have shown that greater specific richness may lead to an increase in the retention of nutrients within the ecosystem. Moreover, modellers have also been able to demonstrate that complexity tends to stabilise ecosystems by dampening the impact of temporary fluctuations in populations. It is similar to a buffer effect.

It has been observed that, in the long term, a certain degree of permanence in an ecosystem tends to promote biological diversity. The great lakes of East Africa (Lake Malawi, Lake Tanganyika and Lake Victoria) are a good example: over the course of their several million years of existence, these lakes have acquired a large diversity of endemic species, including communities of fish and invertebrates that are highly specialized on the ecological level. Conversely, in lake environments of more recent origin, such as those of northern Europe or North America that appeared only after the retreat of the ice caps around 15 000 years ago, communities are not very diversified and are essentially composed of species with a wide distribution range.

The evidence that complexity is important to preserve the entirety and stability of natural systems lends weight to ecologists' argument for the necessity of preserving the totality of the species coexisting in these ecosystems.

5.9   The Role of Biological Diversity in Biochemical Cycles

The productivity of ecosystems depends closely upon the availability of the nutrients controlling the primary production upon which trophic chains are founded. The flow of nutrients is controlled by chemical processes and the biological components in the ecosystems. It is now evident that living organisms play an important role in the dynamics of nutrients. Unrecognized by geochemists for a long time, this role is evidently complex, and we are still a long way from understanding all its implications. It impacts many functions such as nitrification and denitrification, nitrogen fixation, methanogenesis and depollution.

5.9.1 The biological fixation of nitrogen

Nitrogen is the most important constituent in plants after carbon. However, the quantities of nitrogen forms that can be assimilated by plants (ammonium, nitrates, simple organic compounds) found in the soil and waters are often insufficient to ensure the growth of plants. The nitrogen component derives from the biological fixation of molecular nitrogen (N_2), a major constituent of the atmosphere. The latter is a chemically inert gas that can only be used by certain prokaryote micro-organisms called nitrogen-fixing organisms. Thus, the biological fixation of nitrogen is the principal mechanism for introducing nitrogen into the biosphere: in one year, approximately 175 million tonnes of atmospheric nitrogen are reintroduced into the biosphere by micro-organisms, while the amount of nitrate fertilizers used in agriculture totals around 40 million tonnes per year.

In marine environments, only the cyanobacteria have the capacity to use nitrogen molecules to satisfy their metabolic needs. In terrestrial environments, there are two main groups of nitrogen-fixating bacteria that are associated with superior plants:

- the vast group of *Rhizobia*, associated with leguminous plants (the Papillonacea, Mimosacea and Cesalpinacea families);

- *Frankia*, bacteria of the Actinomycetes group, associated with trees and shrubs such as the genera *Alnus*, *Casuarina*, etc.

Nitrobacter bacteria assume the function of nitrification in the soil.

5.9.2 The mineralization of organic matter

Whilst solar energy is continuously dispensed upon the surface of the Earth, the same cannot be said of mineral elements. For life to be maintained, it is therefore necessary for the chemical elements contained in living organisms to be recycled after their death. Prokaryotes play a fundamental role in biochemical cycles: they decompose detritic organic matter, releasing inorganic elements, which then serve to synthesize new organic molecules. If no such decomposers existed, the carbon, nitrogen and other elements essential to life would remain trapped in detritic material.

In the nitrogen cycle, nitrification corresponds to the phase of decomposition of organic matter in an oxygenated milieu. This process culminates in the production of nitrates, which are the form of nitrogen that plants are able to assimilate. When organisms die, their organic matter is decomposed by bacteria, releasing ammonium, of which a certain amount escapes in the form of gas (NH_4^+). In well-aerated soils, nitrifying bacteria can oxidate ammonium: in the first stage (nitrosation), bacteria of the genus *Nitrosomonas* transform the ammonium into the nitrite ion (NO_2^-), which *Nitrobacter* bacteria then transform (nitratation) into the nitrate ion (NO_3^-). This process of *nitrification* provides for a large proportion of the nitrogen assimilated by plants. In anaerobic environments, a reverse process takes place: other kinds of bacteria (see Figure 5.5) transform nitrates into nitrites and gaseous nitrogen. This process is called *denitrification*. Remineralization of nitrogen is thus tightly controlled by a few species of bacteria.

5.9.3   Long-term storage of mineral elements

The biochemical cycles driven by living organisms also lead to the accumulation of large sedimentary formations and consequently to

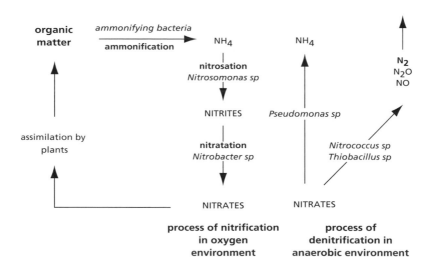

Figure 5.5 Simplified representation of the transformation of organic matter and nitrogen mineralization, as controlled by bacteria under aerobic and anaerobic conditions

the long-term storage of certain mineral elements, often withdrawing them from biochemical cycles for long periods of time.

The large reserve of carbon buried in the Earth's crust in the form of fossil fuels (lignite, coal, petroleum, etc.) derives mainly from the transformation of plants and bacteria that did not go through the process of remineralisation. With spread of plants over the continents in the Carboniferous, large quantities of organic carbon accumulated without being mineralized. At the same time, living organisms acquired the ability to mineralize limestone (molluscs, planktonic organisms such as coccolithophorids). The production of carbonates through biomineralization also contributed actively towards controlling the quantity of carbon present in the atmosphere, whilst a large stock of calcium remained stored in the rocks. The Earth's reserves of carbon sediments and combustible fuels are almost completely isolated from the atmosphere and would be only minimally involved in current cycles and flows, had humans not targeted them as a source of energy.

5.9.4   Recycling and transport of nutrients by consumers

In both aquatic systems and terrestrial environments, there are many examples showing how organisms at higher trophic levels act upon the abundance and dynamics of primary producers. In the process, they may modify the specific composition of the communities, thereby altering the overall functioning of the system. Ecologists have observed that predators can act upon the dynamics of their prey, for example by modifying the quality and quantity of nutrients available to primary producers: the recycling of nutrients via excretion by herbivorous zooplankton and zooplanktivorous fish is an important source of nutrients for phytoplankton in ocean and continental waters.

Due to their mobility, consumers can also transport nutrients to different places within a system. A well-known example is that of Pacific ocean salmon (*Oncorhynchus kisutch*), that return to lay their eggs and die in the upper streams of the rivers after having grown to maturity in the sea. The mass migration of salmon up the relatively unproductive rivers of the west coast of North America constitutes an input of food and nutrients, providing food for various land vertebrates such as brown bears, eagles, otters, etc. The carcasses of the dead salmon are soon invaded by necrophagous invertebrates and large numbers of insect larvae, that serve as food for the young salmon. Analyses of nitrogen isotopes have shown that these carcasses also provide a source of nutrients for river plants and the surrounding, periodically inundated areas.

Recent studies have considerably altered our view of the role of biological diversity in the food cycle. We are only beginning to grasp its complexity in a small number of closely defined situations.

5.10   The Role of Biological Communities

If certain species play a decisive role in ecosystem functioning, whole communities constitute another level of integration within the hierarchy of the living world, fulfilling certain functions.

5.10.1   The importance of micro-organisms for the structure and functioning of pelagic trophic networks in aquatic environments

Despite considerable methodological progress over the last decade, over 90 per cent of the micro-organisms present in the environment have not yet been described. The roles of bacteria and protozoa were once considered negligible in the traditional view of the trophic chain, but we now know that these micro-organisms can exercise significant influence upon the major flows of energy and nutrients.

In pelagic aquatic systems, for example, the somewhat simplistic diagrams of trophic networks conceived in the 1950s were challenged in the 1980s by the discovery of the '*microbial loop*' (Figure 5.6). Parallel to the classic scheme of predation on phytoplankton by herbivores, there exists a microbial trophic network, based on the utilization of carbon deriving from the excretions of living organisms and/or the degradation of dead organic matter by bacteria. These bacteria may be consumed by zooplankton, but their main predators are heterotrophic protists (ciliates, flagellates, amoebae), which are in turn more easily consumed by zooplankton.

A large part of primary production (sometimes over 50 per cent) is thus diverted into the microbial loop through which nutrients are rapidly remineralized and reintegrated into the stock of dissolved inorganic

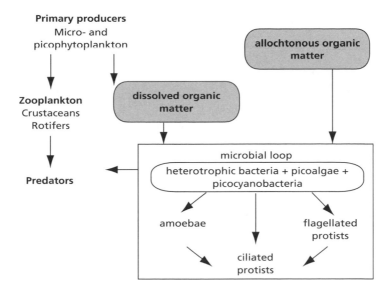

Figure 5.6   The microbial loop in lacustrine environments

substances. Through this process of rapid recycling and remineralisation, the microbial loop ensures the perpetual renewal of nutrients necessary for the growth of phytoplankton.

More recently, it has been demonstrated that viroplankton play an important role in the structure and functioning of aquatic trophic networks. Knowledge of the specific and genetic diversity of viruses is still sketchy, and much progress remains to be made. Their density in aquatic systems is relatively high (between 10^4 and 10^8 ml^{-1}). In functional terms, virus appear to play three major roles:

- as agents for microbial mortality, since they cause cellular lysis. It is currently estimated that viruses are responsible for at least 30 per cent of marine bacterioplankton mortalities;

- as regulatory agents for the microbial diversity in aquatic environments. The functional impact of viruses seems to be particularly significant in processes that are not easily quantifiable, such as the increase of genetic exchanges or maintenance of specific diversity within microbial communities. Some studies indicate that viruses may exert a greater impact upon the genetic structure and specific composition of pelagic algae communities than upon their abundance;

- as recycling agents for organic matter in pelagic environments. Cellular lysis by viruses enriches the environment with released organic matter, which greatly increases the metabolic activity of non-infested planktonic bacteria. There is actually a functional loop linked to bacterial lysis by viruses (bacteria → bacteriophages → released organic matter → bacteria), that contributes to the recycling of nutrients in microbial trophic networks, whilst reducing the contribution of bacterial production to the flow of matter and energy towards higher trophic levels.

5.10.2   Riparian vegetation and the functioning of rivers

Rivers are dynamic systems that interact closely with their immediate terrestrial environment. They erode their banks, periodically flood the surrounding plains, and modify their physiognomy over the long term by creating and destroying dead arms, secondary arms and wetland zones. These spatial and temporal dynamics lie at the origin of the great diversity of plant formations and animal communities in adjacent zones. The specific richness of indigenous riparian vegetation in Europe is comparable to that of certain tropical rainforests. Some 1400 plant species have been recorded on the River Adour in France, with an average of 314 species over a stretch of 500 metres. This riparian vegetation has several functional roles in ecology:

- *stabilization of the riverbanks*: the roots of many trees (willows, alders, etc.) or shrubs form a biological grid that retains sediments and slows the erosion of the riverbanks;

- *flood prevention*: vegetation influences the flow of water; the aboveground parts of grasses, shrubs and bushes reduce the speed of the current and the chance of flooding;

- *creation and diversification of habitats*: branches and dead wood deriving from riparian vegetation were long regarded as an obstacle to the flow of water and a potential risk to riparian activities and civil engineering works. However, we are now aware that coarse wood debris (blockage) plays a role in the ecological equilibrium of rivers by favouring the creation and diversification of habitats. Blockages lead to a succession of rapids and pools, creating microenvironments favourable to the settlement of many species, with a heterogeneity of habitats that enables their cohabitation. Moreover, riparian vegetation provides a temporary habitat for the reproduction, feeding or shelter of numerous land animals (amphibians, birds, mammals). In France, the natural valley of the River Ain harbours 180 bird species (teals, egrets, herons, kingfisher, etc.), of which 100 are nesting species. In general, the large diversity observed in less degraded systems contrasts with the paucity of highly artificial systems such as poplar or conifer plantations;

• *source of organic matter*: riparian woods are a source of allochthonous organic matter (leaves, stems, animals) for the river. It is degraded by the micro-organisms present in the water (fungi, bacteria, etc.). The quality of such inputs varies with the nature and composition of the riparian vegetation: the needles of conifers are far less degradable than the more tender leaves of alders or willows;

• *denitrification and depollution*: through their root system, riparian wood formations also affect the cycle of nutrients and participate in eliminating diffuse agricultural pollution. The vegetation absorbs nitrates and stores them temporarily. Wetlands and wooded environments offer favourable conditions for denitrification by micro-organisms, thereby releasing gaseous nitrogen. By passing through riparian woodland, groundwater is naturally purified of nitrates washed from adjacent agricultural areas.

5.10.3   The role of soil communities

The major interactions among plant roots, animals and microbes that take place under the soil surface directly determine what grows in which place and how.

• One of the basic functions of soil organisms is to *participate in recycling the nutrients contained in organic matter*. The process of decomposition involves both tiny invertebrates that break up organic matter, as well as microbes (bacteria, protozoa, fungi), whose enzymes enable chemical degradation to produce inorganic components that can be reassimilated by plants. The temperature and moisture of the soil affect underground organisms and thus also the rate of decomposition. In a way, the communities of decomposers in the soil are comparable to the microbial loops observed in aquatic habitats.

- Understanding the links between the diversity of organisms above ground and that of organisms below ground is an great challenge for our understanding of how ecological communities and processes are determined at local and regional scales. The factual evidence for a positive correlation is mixed. Across broad gradients in disturbance and other conditions, above-ground and underground diversity may correlate, but the relationship tends to be much more variable at smaller, more local scales. Where correlations do exist, they are not always directly causal in nature, because there are a variety of mechanisms through which organisms above ground could affect community composition and diversity below ground, and *vice versa*. A major difficulty in the way of assessing the links between above and below ground diversity is our scant knowledge of soil organisms: potentially 99 per cent of soil bacteria and nematode species are still unknown.

- One characteristic of underground organisms, as compared to surface communities, is the *importance of mutualist relationships between species*. The majority of dominant plant species in the ecosystems have developed symbioses with microbes living in the soil. Mycorrhizal fungi help plants to find certain nutrients such as phosphorus. In return, they receive carbon from their hosts. In other words, plant species found on the Earth's surface are actually dependent upon underground species in order to function. The specificity of mutualist interactions appears to indicate that the species are not interchangeable, even within the same functional group.

Mycorrhizae

Mycorrhizae are symbiotic associations between plant roots and certain soil fungi. In temperate forests, for example, the mycelia of ectomycorrhizae (chanterelles, boletus, etc.) colonize the exterior of tree roots without penetrating their cells. Mycorrhizae help roots to absorb mineral elements and thus enhance the nutrition of most plant species. They are able to utilize resources which plants would not be able to access on their own (plant detritus and organic matter, insoluble minerals), thereby giving trees access to nutrients. They also contribute towards protecting roots from pathogens. If these fungi are eliminated on an experimental basis, the plants suffer from nutritional deficiencies and perish. In return, the host plant provides its associated fungi with carbon compounds.

5.11 Biological Diversity and the Dynamics of the Biosphere

In the scientific world, compartmentalized into academic disciplines, climate was once the reserved domain of physicists and chemists. But over the past few years, it has been recognized that there is a strong interaction between climate and life: climate affects life on Earth, and conversely, life affects the climate, as well as the global environment in general.

5.11.1 Composition of the atmosphere

Knowledge of the interactions between the biosphere and the atmosphere is essential to understanding climate dynamics. Human activities may demonstrably modify the composition and structure of the atmosphere, with middle-range consequences for the future climate.

A comparison of the Earth's atmosphere with that of Mars and Venus confirms that life plays a decisive role in the gaseous composition of the Earth's atmosphere: the high concentration of oxygen and the low concentration of CO_2 reflect the photosynthetic activity of bacteria, algae and plants over hundreds of millions of years.

The primitive atmosphere of 4 billion years ago contained no free oxygen. According to our present knowledge, its main gaseous components were carbon dioxide (CO_2), carbon monoxide (CO), gaseous nitrogen (N_2), water vapour, and certainly also hydrogen sulphide (H_2S). This atmosphere constituted an anoxic environment; the first living organisms (prokaryote bacteria) would therefore have been anaerobic. Subsequently, the atmosphere grew richer in oxygen. Photosynthesis would have appeared over 3 billion years ago, probably in the form of cyanobacteria capable of fixating both CO_2 and atmospheric nitrogen to synthesize their own chemicals. The transition from an atmosphere without oxygen to an oxygen-rich atmosphere probably took place around

2 billion years ago, but oxygen was slow to accumulate in the atmosphere, and its concentration attained its present value around 700 million years ago. On the one hand, this oxygen was the source for the formation of the ozone layer that protects life from short-wave Sun rays. On the other hand, it allowed new types of metabolisms to appear. The eukaryotes took advantage of the oxygen-poison by developing a metabolism effectively based on oxidation, a considerable source of chemical energy.

These theories attribute considerable importance to the living world: the abundance of oxygen and the low concentration of carbon dioxide in today's atmosphere are the result of photosynthesis.

5.11.2   Control of evapotranspiration in the soil–plant–atmosphere system

The distribution of precipitation, temperature and sunshine largely determines the biological productivity and composition of the plant biomass. In turn, as plant biomass develops, it intervenes in the hydrological cycle.

Where vegetation is abundant, a large proportion of solar energy goes into evaporating the water contained in the plants. Foliar transpiration is a mechanism linked to photosynthesis in plants. A demand for water arises and is transmitted along the stem, causing suction and intake of water and mineral salts from the soil by the roots. The water moves up the stem or trunk to the stomata of the leaves. When the stomata open, water vapour escapes into the atmosphere, whilst CO_2 from the atmosphere penetrates the plant. This water evaporation lowers plant temperature.

Plant transpiration is thus a process for circulating water through a soil–plant–atmosphere system and, as such, constitutes a key element of the hydrological balance. Growing trees create a link between the atmosphere and underground water. The atmosphere above regions covered with vegetation is thus more humid, favouring a nocturnal greenhouse effect that induces precipitation. All this feeds back toward maintaining conditions favourable to plant development. Around half the precipitation on terrestrial ecosystems comes from water recycled by evapotranspiration, a combination of physical evaporation and biological transpiration.

5.12   The Cybernetic Cohesion of Ecosystems: the Role of Communication Networks

Communication between individuals of the same or different species, messages of aggression or cooperation, submission, fear or anger, or messages contributing to the organization of societal life – we now know that the living world is highly structured by a system of communication that went unrecognized for a long time. While there remains much to explore in this realm, the results already suggest that communities are not simply chance collections of juxtaposed species. Interactions between species, whether belonging to the same or different taxonomic groups, are expressed through a wide range of signals that make it possible to preserve the cohesion of populations or communities and/or modulate the behaviour and biology of species.

In light of new discoveries, it appears increasingly likely that there exists a communication network superimposed upon the network of matter and energy flows. This could be described as a form of *cybernetic cohesion* based on the communication network linking the biotic components of an ecosystem.

In aquatic systems, a large variety of signals are used within and among species:

- *visual and light signals* are used in waters with enough light for such signals to be perceived. The mating rites of many varieties of fish combine movement, colour and the relative positions of the partners;

- *chemical signals* make use of a large variety of substances; however, they are relatively slow and non-directional. Communication via chemical mediators is believed to have been the earliest form of exchanges between individuals and species;

- *electrical signals*: certain fish species, in particular those belonging to the Mormyridae, use a rich repertoire of electric signals. They make it possible for these species, who generally live in turbid waters, to maintain elaborate social relations;

- *acoustic signals and vibrations* are used by vertebrates (fish as well as aquatic mammals such as whales) and invertebrates. Sounds travel faster in water (1500 m s^{-1}) than in the air and are therefore a good means of communication over distances where vision is impossible.

Aquatic systems actually appear to be far noisier than previously believed.

5.12.1 Communication by chemical substances in aquatic systems

The past few years have seen considerable progress towards understanding the ways in which aquatic organisms communicate via chemical substances. At least four major types of signals are used:

- *alarm substances* are produced by the tissues of wounded fish and alert other members of the species, inciting reactions of fear. Similar phenomena have been observed for invertebrates;

- *repellants* enable prey to ward off predators. Many kinds of algae have a toxic effect upon zooplankton, and zooplankton is apparently able to recognise toxic algae and avoid ingesting them;

- *kairomones* produced by predators elicit reactions in their prey, provoking modifications in their behaviour, morphology and biological characteristics. Numerous studies have demonstrated that chemical substances emitted by predators (fish, *Chaoborus* larvae) affect the morphology as well as the adult size and fecundity of crustacean plankton belonging to the *Daphnia* genera;

- Both vertebrates and invertebrates use a great variety of *pheromones* to help recognize sexual partners.

Among terrestrial plants and animals, individuals can communicate in different ways.

- Mediating organisms such as mycorrhizal fungi can establish liaisons between the root systems of plants, placing them in communication with one another. Individuals of the same or different species can be interconnected in this way. Both cases are known to involve a transmission of substances (mineral nutrients, carbohydrates).

- Chemical elements may be transferred from one plant to another through air or water. Plants cannot flee to escape from aggressors, but that does not mean that they remain passive. According to recent studies, volatile substances of vegetal origin can trigger defensive mechanisms in healthy plants situated near infested plants. It has also been demonstrated, both in the laboratory and in agricultural systems, that an attack by herbivores increases the emission of volatile organic compounds which attract predators of the herbivores to the infested plants. In other words, the plants mobilize their 'lifeguards' to protect themselves against their enemies. Wild tobacco (*Nicotiana attenuata*) manages to rid itself of 90% of its herbivore predators in this way. It still remains to be understood in detail why and how such volatile substances can be emitted by plants under stress and how healthy plants utilize this information, but the phenomenon as such appears to be well established. Given these circumstances, it is legitimate to consider whether plants might not be able to anticipate certain risks thanks to signals informing their recipient of danger in the vicinity.

Scientists are already exploring what role man-made chemical substances and their residual products play for the cybernetic coherence of ecosystems. Could artificial compounds imitate natural substances (mimetic compounds) and influence the behaviour of species?

6 The Dynamics of Biological Diversity and the Consequences of Human Activities

Biological diversity is not the result of a uniform process. It is the heteroclitic heritage of living beings whose main characteristics were established several hundreds of thousands of years ago. Certain species and lines have died out, many of them in conjunction with the great crises that have punctuated the history of the planet Earth. Of others only a few remaining descendents survive under particular conditions (the coelacanth, for example), while others have diversified broadly. Certain types of ecosystems have been severely impacted by climatic changes over the last few million years. A cycle of ice ages alternating with global warming has acted as a kind of ice-scraper, causing the flora and fauna of northern Europe, Asia and North America to disappear almost entirely. The development of studies on paleoenvironments enables predictions about future climate changes. This also helps to explain current population assemblages in ecosystems. The hope is that knowledge of the past will allow us to anticipate the future and orientate our actions accordingly.

In geographic terms, the distribution of biological diversity is the result of both present climatic conditions, and the way they shape large biomes and regulate ecosystem functioning, as well as past fluctuations, whether they be geological (the origin of the continents, orogenesis) or climatic changes that enabled or disenabled the survival and evolution of particular species (Figure 6.1).

The future of biological diversity depends upon the same factors that have created the current situation, with an added factor: humankind, a

Biodiversity   Christian Lévêque and Jean-Claude Mounolou
© 2004 John Wiley & Sons, Ltd   ISBN 0 470 84956 8 (Hbk)   ISBN 0 470 84957 6 (pbk)

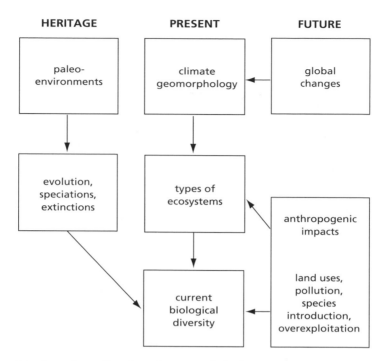

Figure 6.1 In a given climatic and geomorphologic context, the current biological diversity is the heritage of evolution and environmental history, a heritage that incorporates both the dynamics of climatic variations and human activities

recent invasive species, modifies its environment over space and time on an unprecedented scale, thereby menacing a proportion of biological heritage and the survival of numerous other species either directly or by modifying the conditions for their existence. As a result, some say that we are currently experiencing a sixth period of mass extinction.

6.1   Paleo-environments and Biological Diversity

Landscapes change slowly, as long as they are not subject to serious impacts due to human activities. Measured on the scale of a human life, this may give us the impression that the world around us is stable. Our sense of a 'natural equilibrium' leads us to believe that there is a fixed *status quo* for nature, a standard condition. Only human activities would modify them in the short term. A simplistic concept of conservation takes this 'Garden of Eden' as a point of reference and strives to preserve it from harm or repair damage done, so as to re-establish its pristine state. In reality, this approach disregards a basic axiom: biological diversity is in a state of perpetual evolution on scales of time and space that do not necessarily coincide with human scales. Biological diversity has a history, a present and a future that we must understand if we are to adopt adequate measures towards its conservation.

The Quaternary, that began less than 2 million years ago, offers a good illustration of the processes involved in the dynamics of biological diversity. It is marked by the occurrence of several ice ages and extreme variations in climate. Scientists have been able to reconstruct a relatively accurate picture of the dynamics of ecosystems and biological diversity since the last ice age cycle as a function of climatic changes over time. They have also demonstrated that human activity has been impacting environments and species for a long time, albeit not to the extent observed today.

6.1.1 Northern European terrestrial systems

During the last ice age, which peaked around 18 000 years ago, a vast sheath of ice once again stretched over northern Europe, and an immense glacier covered the Alps. Sea level was 120 m below its present level. In France, the mean temperature was 4.5 °C less than it is now, and all that survived was a blanket of herbaceous plants extending from the icy regions in the north to the Mediterranean in the south. In these subarctic steppe and tundra landscapes, similar to those of Lapland today, trees had become extinct (summer temperatures were too low). Reindeer, horses and bison, as well as extinct species such as the mammoth, cave bear and *Megaceros*, populated the southerly regions of the territory. Altogether, the species and ecosystems were very different from those we know today.

During this same, most recent ice age, ligneous plants that were relatively thermophile survived in southern Europe in the lower parts of rivers and on the southern slopes of the mountains at the periphery of the Mediterranean region. A number of species such as oaks (see box) survived in three refuge zones situated on the Iberian and Italian peninsulas and in the Balkans (Figure 6.2). Other refuge zones existed in eastern Europe and in Asia. In all cases, however, the assemblages of animal and plant species that survived were fragmentary and to a certain extent chance assemblages.

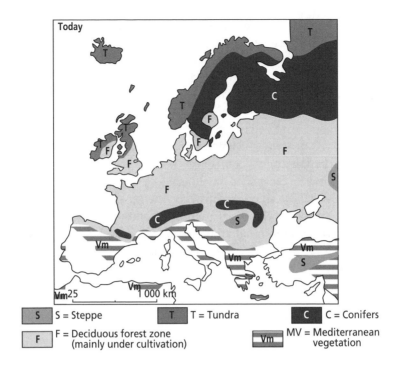

S = Steppe   T = Tundra   C = Conifers

F = Deciduous forest zone (mainly under cultivation)   MV = Mediterranean vegetation

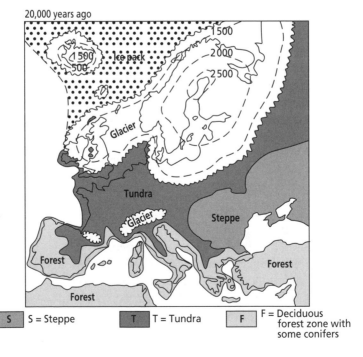

S = Steppe   T = Tundra   F = Deciduous forest zone with some conifers

During the warming period that followed, the biotope was in constant evolution as the plant and animal species that had survived in refuge zones recolonized periglacial regions. Around 16 000 years ago, there was a phase in which steppe-like herbaceous formations extended over the whole surface. During the period of transition between the last ice age and the Holocene period, at the medium latitudes of northern Europe, juniper (*Juniperus*), followed by birch (*Betula*) and pine (*Pinus*) began to recolonize regions abandoned by the forest during the ice age. Some 10 000 years ago, temperate deciduous forests started propagating over all of Europe. Oak (*quercus*), a tree that competes well with other species, colonized Europe more rapidly than the others. It was accompanied by hazel (*Corylux*) and elm (*Ulmus*). Later, ash (*Fraxinus*) and lime (*Tilia*) appeared in mixed forests dominated by hazel and oak. The spread of beech (*Fagus*) from its refuge zones near the Black Sea and in southern Italy was delayed, because it had difficulty competing with oak. It began propagating throughout the Apennine mountains 6000–6500 years ago, and it was only 3500 years ago that it appeared in Spain and in northwest Europe. As for spruce: it was prevalent in the eastern Alps after the last glaciation, took over 6000 more years to reach the French Alps and the Jura Mountains and only colonized the Massif Central in the 19th century as a result of human reforestation activities.

Generally speaking, the dynamics of plant succession were similar all across Europe, although species appeared with the varying synchronies, depending upon the geographic situation. Tree recolonisation was simultaneously a function of climate, the refuge zone situation, and the competition among species within the ecosystem undergoing restructuration. For sessile plant species, an important factor is their speed of propagation, which depends upon the biological characteristics of the species.

Figure 6.2 Europe as it was 20 000 years ago (right-hand page) and recently, prior to ground-clearing by humans (above)

Twenty thousand years ago, one large ice cap extended over the British Isles and Scandinavia. A glacier also covered Iceland, with pack ice joining the two land masses. Vegetation characteristic of cold climates covered most of Europe. Steppe (herbaceous vegetation) covered the east, and tundra (herbaceous and shrubby vegetation) the north and west. Trees were only encountered in the extreme south. When the sea level dropped, the continents extended, linking France and Britain. The second map illustrates a recent state, with tundra high in the north and steppe far to the east. Most of Europe is covered with forests (adapted from Foucault, 1993)

The pedunculate oak was able to recolonize Europe relatively quickly, between 13 000 and 6000 years ago, at an average pace of 500 m per year. Such rapid progress for a non-motile organism with a slow growth rate (an oak is not fertile before the age of 15) was facilitated by jays, which transport the acorns and store them as reserves, burying them sometimes 4–5 cm deep at a distance of several kilometres. A jay may hide around 5000 acorns per year in sites with sparse vegetation. It is estimated that half the fruits dispersed by jays may yield a shoot, making this species responsible for over half of the natural regeneration of oaks.

In addition to natural propagation out of the refuge zones, other trees also benefited from human transportation. Thus, for example, the plane tree made its way from Greece to western Europe in the baggage of the Romans; walnut was transported from the Balkans, and cypress from Asia Minor – all the work of the Romans.

Recent studies have traced the recolonization routes of flora and fauna after the ice ages. For species such as the brown bear and rabbit, the route runs from the Iberian peninsula to Great Britain and all the way to southern Scandinavia. The barrier of the Alps would have delayed or prevented the dispersal of many of the species that had found refuge on the Italian peninsula. Joint comparison of these colonization paths points up the existence of so-called 'suture' zones – zones of hybridization between populations of the same species that met as they spread out from different refuge zones to recolonize Europe (Figure 6.3).

In general, relatively poor biomes reinstated themselves in temperate and cold regions, while much richer, Mediterranean forest groups were reconstituted in the South. The alternation between ice ages and intermediary periods has recurred approximately every 100 000 years over millions of years, with varying degrees of intensity. These cycles have caused frequent and profound changes in the flora and fauna of northern Europe, which have been far from stable over these periods.

6.1.2   Tropical rainforests

Discussions on tropical forests tend to be somewhat irrational, focusing upon the question of protecting these ecosystems, which are undoubtedly severely disturbed by human activities. The common assertion that tropical forests are the lungs of the Earth proceeds either from erroneous

European oaks and climatic cycles

The diversity of European oaks is the result of a succession of ice age periods alternating with climatic warming during the Quarternary. These climate changes caused cycles of extinction followed by recolonisation. In the process, certain species were selected, while others were eradicated. Since the last ice age 18 000 years ago, oak forests have repopulated the continent. A major European study inventoried the dominant species and decoded their colonization strategies. Chloroplast DNA analysis showed that during the last ice age, oak populations were confined to three refuge zones in the south of the European continent. When the climate began to warm 15 000 years ago, these populations, which had remained separated for 100 000 years, embarked upon the reconquest of territories to the North, entirely covering their current domain 6000 years ago. On an average, the oaks advanced 380 m a year, reaching peaks of 500 m during certain periods. In addition to continuous dispersion, a mechanism for discontinuous dispersion by successive jumps (with rare occurrences of long-distance dispersion over several dozen kilometres) explains the rapidity of this progression.

Nuclear DNA analysis also revealed a remarkable strategy implemented by the oaks to promote their dissemination. Each of the four main species of European oak (the sessile oak, the pedunculate oak, the Pyrenean oak and the downy oak) has its own ecological preferences, whether for forest environments or open environments, humid soil or calcareous soil. There was actually a kind of 'teamwork' between these species, co-operating to conquer new environments. One species would assume the role of pioneers for the benefit of the others. The mechanisms also include extensive genetic mixing among the species and interspecific hybridization.

science or from a conscious contempt for scientific results. In fact, the tropical forest that the collective unconsciousness so often depicts as an example of wild and pristine nature has actually experienced periods of expansion and retraction in connection with pronounced climatic changes.

During the last ice age, under the impact of most severe climatic conditions (low temperatures of 2° to 6 °C and little precipitation), the South American rainforests began to recede 28 000 years ago. The

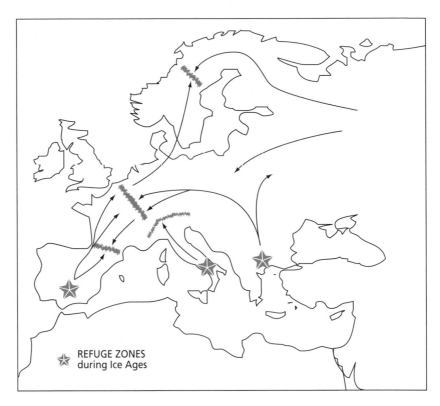

Figure 6.3 Principal refuge zones and post-glacial recolonization routes in Europe (adapted from Taberlet *et al.,* 1998)

African forests followed suit between 20 000 and 15 000 thousand years ago. In many regions, grasslands and/or savannas supplanted the rainforests, which continued to occupy only limited surfaces.

Since the postglacial period, the history of forest re-expansion has not been monotonous. In Africa, there is evidence that forest coverage reached its maximum between 10 000 and 8000 years ago and maintained that extent until 5000 years ago (Figure 6.4). After this period, there was less rainfall in central Africa, and between 3000 and 2500 years ago, the grasslands advanced to the detriment of the forest in the southern Congo and in other regions where the seasonal droughts were the most extreme. Severe disturbances also occurred in the rainforest of western Cameroon. But for the last millennium, there has been a general tendency for forests in central Africa to expand once again, advancing by hundreds of metres

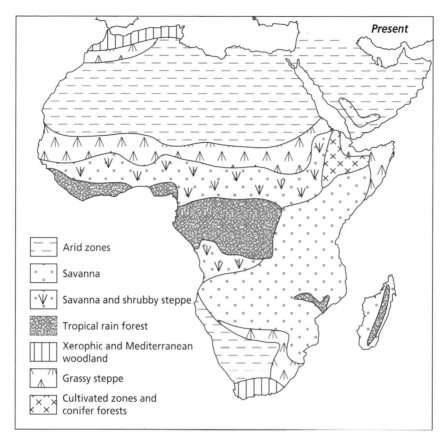

Figure 6.4 (a) Schematic map illustrating the changes in the distribution of the principal domains of vegetation across Africa since the last Ice Age

per century. This forest re-expansion is probably linked to the return of wetter climate conditions.

In South America, the Amazon forest has experienced severe disturbances since the last ice age, but forest dynamics are not in synchrony with those in Africa. From 10 000 to 8000 years ago, the forest developed in some places (southwestern Amazonia and in central Brazil), but not in Guyana. Between 7000 and 4000 years ago, the forest receded, making way for herbaceous formations in northern and southwestern Amazonia, as well as in central Brazil. Beginning 4000 years ago, the forest has been regaining sites where it had disappeared, reaching its full extent on the

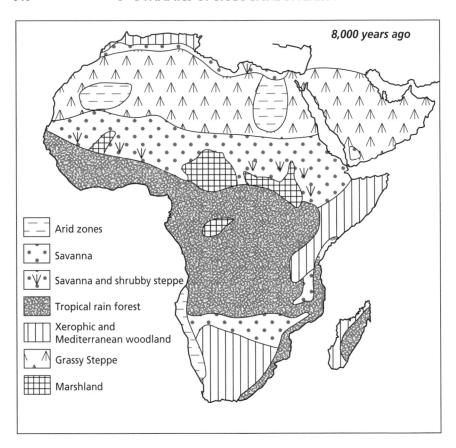

Figure 6.4 (*continued*)   (b) Distribution of the principal vegetation biomes across Africa 8000 years ago

Atlantic coast of Brazil only a thousand years ago. In other words, the Amazon forest is relatively young!

Studies in Guyana, on the other hand, have shown that the forest was more humid between 3000 and 2000 years ago than it is now. Between 1700 years ago and today, there were two episodes of drought linked to disturbances of the forest system, creating large forest clearings that were favourable to the development of pioneer plants: one episode between 1700 and 1200 years ago, and another between 900 and 600 years ago. Comparable phenomena occurred during the same periods in the western Amazon basin. The Guyanan forest acquired its present characteristics only 300 years ago.

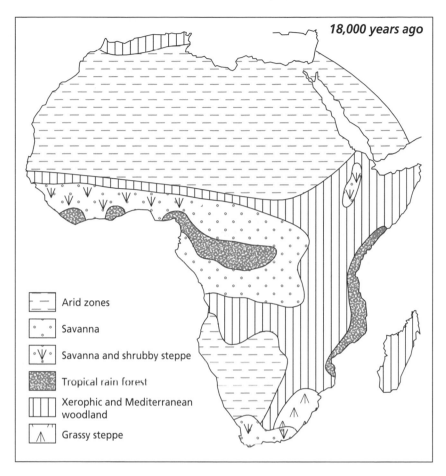

Figure 6.4 (*continued*) (c) Distribution of the principal vegetation biomes across Africa 18 000 years ago

Fire is an important element of disturbance in tropical rainforests. At present, the constant high level of humidity does not allow fires to develop over the whole of such ecosystems,. However, it has been shown that in Amazonia and Guyana, fires and large-scale forest destruction have recurred at intervals up until relatively recent times. Particularly in Guyana, the paleo fires known to have occurred between 8000 and 6000 years ago, and again from 4000 to 2000 years ago, were probably of natural origin, since there is no evidence of human occupation until 2100 years ago.

The hypothesis of refuge zones

The hypothesis of refuge zones stemmed from observation of the current distribution of plant and animal species in dense rainforests and the levels of differentiation attained. In dry periods, the forest is fragmented, with only scattered patches subsisting in zones with amenable climatic conditions. Those forest patches that survived the dry periods served as refuges for a proportion of the species that previously occupied the territory covered by the forest. If the fragmentation lasted long enough, it would foster speciation at second hand. The species would then have reinvaded the zones in between the refuges at variable speeds, depending upon their ability to migrate, which depends in turn upon their biological characteristics. According to the theory, the former refuges should correspond to current zones with high endemicity and large biodiversity, with diversity decreasing as the distance from refuge zones increases.

The comparison of tropical forest flora reveals profound differences between the three continents (America, Africa and Asia). These are partly attributable to their different climatic histories. Malaysia's flora, for example, is twice as diversified as that of Guyana and richer in endemic species. The two sets of flora also exhibit different biological characteristics regarding their modes of dispersal and their rates of endemicity. In Guyana as in other forest zones of tropical America, around 80 per cent of the tree species are dispersed by mammals and birds. The figures are comparable for Africa. In tropical Asia, on the other hand, the majority of tree species produce dry fruit whose grains fall to the ground and remain at short distance from the fruit-bearing tree. There is no system of rapid dispersal. The differences observed between Malaysia and Guyana may also be explained by the longer periods of dryness in Malaysia. Moreover, the mountainous topography of Malaysia provides more possibilities for refuge zones, alleviating the impact of climatic disturbances. The relatively poor species richness of Guyana as compared to Malaysia is probably the result of extreme forest recessions during dry climatic phases as well as its fewer and smaller refuges, owing to its more monotonous surface structure.

6.1.3   Continental aquatic systems

Generally speaking, the extent of aquatic habitats fluctuates as a function of climatic conditions and over relatively short periods on a geological scale. With some exceptions, continental aquatic ecosystems can be characterized as transitional systems in that they are continuously reacting to climatic variations. In temperate regions, ice ages have, at various intervals, caused the disappearance of aquatic environments and the extinction of the associated fauna. This explains why the species richness of fish populations is lower in temperate regions than in equatorial regions. It is interesting to compare Lake Geneva, which was covered by glaciers at the height of the last ice age (20 000 years ago), with the great lakes of east Africa (Lake Tanganyika, Lake Malawi), which are known to have existed for several million years. Lake Geneva is a young lake whose present fauna is the result of recent recolonisation following the Holocene warming period. The colonizing species came from refuge zones where at least a portion of the aquatic fauna could survive. The fish fauna is relatively poor, comprising only 14 indigenous species. By contrast, east African lakes are ancient lakes, existing continuously over several million years, even if their water levels varied by several hundred metres during this period. These lakes are home to a highly diversified ichthyic and invertebrate fauna, the result of a long coevolution of the environment and species. In schematic terms, the fauna of these lakes diversified so as to optimise exploitation of all the resources of the ecosystem. Lake Geneva, by contrast, is populated by a set of species that is still heteroclitic. It is clearly impossible to understand the biological functioning of these lakes (production, trophic networks) without understanding their history. Studying the east African lakes, it is possible to evolve theories on coevolution and speciation, adaptive radiation, concepts of niches and competition for resources. In Lake Geneva, it is merely possible to ascertain that a certain number of species have succeeded in recolonizing the lake since the ice disappeared; however, the situation, which is partially based on chance, does not lend itself to the application or development of theories on evolution.

The situation is similar for river systems. Climatic variations could lead to significant changes in their morphology and sometimes even cause their temporary disappearance. For fish to repopulate basins that have dried out or frozen over, or to colonize new watersheds created by geological and climatic events, physical communication with the refuge zones that preserved a diversified fauna is necessary. For example, the fish fauna of Ireland is extremely poor. At present, it comprises around twenty species, many of which were reintroduced by humans. Indeed, after the glaciers receded in the Holocene, only eight anadromous species (salmon, shad, eel, etc.) were able to recolonize Irish waters by natural routes. The situation in North America is quite different: the Mississippi river, which was the main refuge zone for aquatic species during recent ice ages, runs from north to south. Therefore, as the glaciers advanced from the north, species could migrate south and take refuge in the southerly part of the river, moving upstream again as the climate grew warmer. The Danube was a refuge zone for fish in Europe. But since it runs in from west to east, species did not have the possibility of migrating as far south as in North America. Thus, there was probably a far greater number of extinctions during the ice ages, which would explain the relatively depauperate fish fauna in Europe as compared to North America.

6.2   Humankind and the Erosion of Biological Diversity

It is generally held that humans are responsible for a current major phase of extinctions of biological diversity. This assertion is not fundamentally wrong, but for the moment, it is difficult to assess the amplitude of the phenomenon. Is its impact greater than those of climatic disturbances or those attributable to El Niño, which causes droughts and floods, coral bleaching, etc.?

Rather than consider the consequences of human activities globally, it is more helpful to analyse their different effects on various animal and plant groups in their regional and local context. It is quite likely, as a matter of fact, that the current extinction of species is caused by the synergy of several factors.

6.2.1   The myth of the noble savage

Ethnologists have lent some support to the idea that so-called primitive societies (as compared with industrialized societies) had little impact upon the natural environment and lived in a kind of equilibrium with their environment. This myth of a Garden of Eden has proved particularly enduring in the debate that has evolved around traditional knowledge and has been adopted by certain conservationist movements and a proportion of the general public. However, a series of observations tends to demonstrate that modern humans are merely continuing a process of biodiversity erosion that began a very long time ago, at the time of our distant ancestors.

Early explorers regarded the islands of the Pacific as true paradise. Their reports contributed to the myth of primitive societies living in a 'state of equilibrium' with their environment. These regions were populated by human beings at a late date: e.g. the Fiji and Tonga islands 3500 years ago; New Zealand 3200 years ago; and the Hawaiian archipelago 2600 years ago. As for the Galapagos Islands, they were not inhabited until the arrival of Europeans in 1535. Today, we know that one-third to one-half of the bird species that inhabited the Pacific islands has disappeared since human colonization. This represents some 10–20 per cent of the terrestrial bird species known today.

It has been shown that in New Zealand, almost all the bird species of the later Pleistocene period survived changes in climate and vegetation until the arrival of the Maori 800 years ago. Over one-third of its terrestrial birds have disappeared since this time. On the islands of Hawaii, almost two-thirds of the species disappeared over a few thousand years. There are multiple reasons for these extinctions: first, hunting, as well as egg collecting, facilitated by the existence of many running birds incapable of flying; second, the introduction of domestic or commensal animals (dogs, pigs, rats, etc., as well as cats, which arrived with the Europeans) contributed greatly to the extermination of small bird species. However, it is also known that humans profoundly altered the vegetation. Thus, half of New Zealand's forests were destroyed by the Maori from 800 to 500 years ago to make room for fern-covered areas or grasslands. The analysis of fossilized pollens shows that Easter Island was once covered with a luscious subtropical forest. All that remains today is an impoverished prairie, and the largest indigenous animals are insects.

Two conclusions can be drawn from these observations. On the one hand, it is undoubtedly clear that the successive human colonizations of the Pacific islands led to the disappearance of many vertebrate species. This is of relevance to the question of the human impact upon natural environments over the course of millennia. On the other hand, biogeography must also take into consideration the apparent changes in the distribution of species today as well as that of extinct species, to the extent that it is possible to reconstitute past settlements on the basis of relevant fossils.

6.2.2   The extinction of large mammals at the end of the Pleistocene: were humans the cause?

In addition to the massive and spectacular extinctions, more limited extinctions occurred, affecting only one or some groups of organisms. Thus, over the last 50 000 years, hundreds of terrestrial vertebrates have disappeared from different continents and a number of large islands, without being replaced by other species. These extinctions have mainly affected the megafauna. Only a few small species have been impacted, while all varieties of mammals weighing more than a tonne, as well as 75 per cent of varieties between 100 kg and 1 t, disappeared from most continents, with the exception of Africa. The question is whether these extinctions resulted from the expansion of the human species, or if there were other causes.

In fact, the extinctions did not occur on all continents in the same period. In Australia, colonized by *Homo sapiens* 55 000 years ago, all large and medium-sized mammals disappeared around 50 000 years ago. All species heavier than 100 kg and 22 of the 38 species weighing between 10 and 100 kg disappeared, as well as three large reptiles and the giant emu *Genyornis*, which weighed over 200 kg.

Some 12 000 years ago, North America was home to a spectacular megafauna, including three kinds of elephants, three kinds of cheetahs, many different kinds of antelopes, camels, uamas, horses, bison, tapirs, giant wolves, etc. There were more large animals than currently exist in Africa. And yet, around 11 000 years ago, almost all these large animals (70 species, or 95 per cent of the megafauna) disappeared completely. This corresponds to the time when North America was colonized by the human species, and some scientists see archaeological proof that this extinction was caused by hunting. South America was also colonized by humans 11 000 years ago and has since lost 80 per cent of its large mammal varieties.

In Eurasia, the large mammal fauna consisted of animals adapted to the cold: the woolly rhinoceros, the cave bear, joined by animals adapted to temperate periods when Europe was covered with forests – stags and deer, wild boars. A large proportion of this fauna also disappeared in the time between 12 000 and 10 000 years ago.

The situation is different in Africa, although that is where humans evolved over millions of years. The megafauna is still well represented, even if 50 varieties disappeared around 40 000 years ago. Africa is the continent that currently has the most diversified fauna of large herbivores, including the elephant, hippopotamus, rhinoceros, etc. – groups that were plentiful on other continents before the Pleistocene extinctions.

The situation is paradoxical: in Africa, where humans have existed for the longest time, the megafauna is more varied than on continents more recently colonized by humans. On the basis of the information at our disposal, it is difficult to confirm (as some are nonetheless quick to assert) that humans are responsible for the extinction of megafauna at the end of the Pleistocene period. Most likely, its disappearance on the different continents was due to a set of factors acting more or less in synergy. Some of these factors could be as follows:

- Climate changes at the end of the Quaternary, when ice ages alternated with warming periods approximately every 100 000 years, probably played a significant role in modifying habitats and weakening the flora and fauna.

- Hunting could have played a role, at least for certain species and on certain continents, although ground-clearing by fire and the introduction of domestic or commensal animals are probably also factors in the disappearance of large mammals.

- Epidemics could have brought about the extinction of at least some of the species, particularly where the hosts came into contact with new pathogens.

6.2.3 The current erosion of biological diversity

Modern humans are endowed with unequalled technological means. We can make certain ecosystems disappear altogether or transform entire regions. For well-documented groups such as mammals and birds, or certain plant groups, registered extinctions provide a good overall indication. Accordingly, an estimated 108 bird species and 90 mammal species have disappeared since the year 1600. However, the real number is probably higher, because not all regions of the world maintain archives for reference. Moreover, the populations of many species have fallen to critical levels at the present time.

A large proportion of the extinct species, whether they be mammals, birds, reptiles, terrestrial molluscs or plants, previously inhabited islands. One such example is the famous dodo that lived on the island of Mauritius. But continental species such as the aurochs, the American passenger pigeon or the Emperor penguin have also been exterminated by hunting. In the marine domain, we know of only two mammalian species that have disappeared in recent centuries, although certain whale populations have experienced critical periods.

Non-governmental organizations such as the IUCN (The World Conservation Union) have compiled 'red lists' of extinct and endangered species. Altogether, 584 extinct plant species have been recorded and 641 animal species, or an average of three extinctions per year since the beginning of the 17th century. These figures may appear modest, but they are, by all appearances, biased, since many extinctions have gone unrecorded and we are a long way from knowing all species.

Another approach takes account of the fact that not all species are known. It is mainly based on the relationship between the surface area covered by a type of ecosystem and the number of species that can live there. This is the well-known area–species relationship of biogeographers. For example, knowing the reduction in the surface area of the tropical forests, it is possible to predict the number of species that will disappear: between 25 000 and 50 000 per year, according to the estimates. These extinctions mainly include unrecorded arthropods that would vanish as they lived, with complete anonymity. Questionable as the results obtained by this method may be, they support the thesis that human activities are responsible for the current crisis. They are also used by forecasters to assert, in the absence of other proofs, that one-quarter or half of biological diversity will disappear between the present and the end of the 21st century.

In reality, for many groups of flora and fauna, there is a lack of reliable data on the number of living species and the species presumed extinct (see Table 6.1). Given this context, it is difficult to offer serious quantitative information beyond that pertaining to a few limited taxa. There is, of course, no question of claiming that humans have no impact upon the living world; only that this impact is probably not identical for all groups under consideration. Clearly, some catastrophic assertions are founded more upon personal conviction or a desire for publicity than upon science.

Some of these evaluations of the erosion of biological diversity are quantitatively debatable; moreover, they do not take phenomena of

Table 6.1   The evolution of population numbers of vertebrate fauna in France

	Number of species at the beginning of the 20th century	Number of extinct species	Newly appearing species	Introduced and acclimatized species	Number of species in France in 2000
Freshwater fish	69	−3	+2	9	77
Amphibians	35	0	0	3	38
Reptiles	37	0	0	1	38
Birds	352	−4	+4	5	357
Mammals	116	−3	0	7	120
Total	609	−10	+6	25	630

speciation in to account. Just as species evolve and adapt to natural environmental changes (this is what drives biodiversity), it is conceivable that species also evolve under the impact of man-made disturbances. One of the main factors responsible for speciation is allopatry – the geographic isolation of populations that continue to evolve independently of one another. The construction of reservoirs, for example, isolates animal and plant populations, enabling each to evolve separately on their own side of the dam. Introducing species to different, separate regions of the world is another way of creating the conditions for allopatric evolution. The time it takes for species to evolve depends upon the groups under consideration, but little is actually known about the pace of speciation for most groups.

Still less is known about micro-organisms, but the evidence indicates a situation very different from that of vertebrates. Micro-organisms evolve very rapidly and adapt quite well to new conditions created by humans, as evidenced for example, by their resistance to antibiotics, pesticides, etc. (see Chapter 3). The evolution of certain viruses is estimated to be two million times faster than that of animals. All signs point to the conclusion that human activities are actually increasing microbial diversity.

Cities and biological diversity

Cities are human creations. Big cities occupy large areas of the planet and have supplanted vast rural spaces. While the growth of urban environments has contributed towards the disappearance of numerous species, it has also provided certain species with new opportunities for

continues overleaf

Cities and biological diversity (*continued*)

colonisation. The diversity of species may even be greater in the city than in some rural habitats subjected to intensive farming. The fact that cities have expanded on a large scale only recently has not given species much time to evolve. But by adapting their behaviour, some colonizers have been able to make the most of the new conditions offered to them: the presence of many vertical structures favourable to colonization by plants and animals; a more temperate climate; new food sources; a limited number of predators, etc. Thus, the fox population in central London is probably approaching several thousand individuals. Wood pigeons have found refuge in the cities, where they are not hunted, and since the 1960s, turtle-doves have made their home in large cities, where they compete with the common pigeon and wood pigeon. Intercontinental translocation of species has led to a diversification of urban biological diversity: different parrot species have populated the parks of Europe's capital cities, in addition to various invertebrates that originally arrived in containers.

6.3 Human Activities and the Dynamics of Biological Diversity

Human population growth and the corresponding demand for natural resources on the one hand, and the development of industrial, agricultural and commercial activities on the other, are transforming the surface of the Planet and modifying biogeochemical cycles, along with the biodiversity composition in most terrestrial and aquatic ecosystems. Such impacts are reasonably well known and quantified. They in turn cause retroactive changes in the functioning of the biosphere, ushering in climatic evolutions and leading to an irreversible loss of components of biodiversity (genes, species, ecosystems). These influences of humankind upon the biosphere are manifest in different processes, as summarized in Figure 6.5.

The expression 'global change' is often used to characterize all these phenomena, which fall into four categories:

- changes in land use and vegetation;

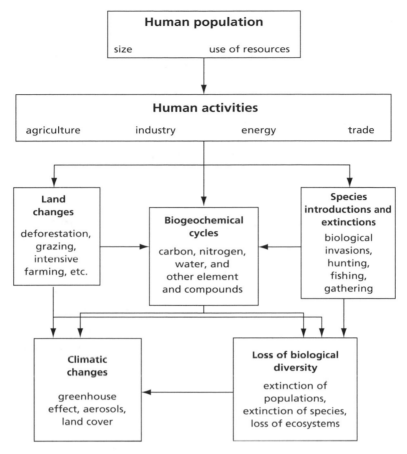

Figure 6.5 Conceptual model illustrating direct and indirect effects upon the biosphere (adapted from Lubchenko *et al.*, 1991)

- changes in atmospheric composition;
- climate changes;
- alterations in the composition of natural communities and loss of biodiversity.

The term implies the impacts of both economic development and world population growth upon the principal domains of the Earth's system – the atmosphere, soils and aquatic systems – as well as those processes involved in the exchanges of matter and energy between these domains.

6.3.1   Demographic pressures

Among the factors responsible for the erosion of biological diversity, demographic pressures and increasingly powerful technologies are the ultimate causes. More space is required to house and feed a world population that has increased dramatically: 2 billion people in 1930, 4 billion in 1975, and predicted to reach 8 billion or more as we approach 2020. This population growth is of concern to the whole planet, but particularly for tropical countries, where biological diversity is greater than in temperate zones.

6.3.2   Land use and the transformation of the countryside

In order to produce goods and services, humans modify the structure and functioning of ecosystems. The first major actions of humans upon their environment involved setting fires in order to flush out wild game or clear the ground. Fires favoured resistant plant species and the development of savannas and grasslands. Next, the emergence of agriculture initiated a transformation process of ecosystems in which domestic species dominated along with opportunistic species. Hedgerows created enclosures for animals, whilst at the same time preserving useful tree species. These enclosures are an essential feature of the contemporary landscape in Europe, where the structure of plant communities is largely dependent upon humans. In other words, that which we call 'Nature' is actually the result of thousands of years of land use by human societies who moulded the landscape.

Some systems have been completely modified by humans: on the global level, 10–15 per cent of the land is used for farming, and 6–8 per cent has been converted into pastures. The other ecosystems, i.e. the majority, are only used for hunting or gathering or for the more or less industrialised extraction of biological resources (timber, fish). Altogether, the proportion of land that has been transformed or degraded by humans is estimated between 40 and 50 per cent. These transformations are the main reason for the erosion of biological diversity.

On a global scale, forests are losing ground. The situation is particularly worrying for tropical forests, but it is also beginning to get very serious for boreal and temperate forests. France, however, is witnessing the reverse phenomenon: over an interval of 40 years, woodlands have grown from 11.3 to 15 million hectares, to the point where forests now occupy more than one-quarter of the country's territory. This is partly the result of a voluntary policy as well as the creation of the ONF (National Forest Office) in 1966, which manages over one-quarter of the national forest land. But the overall figures conceal profound changes:

- Forested surfaces are extending at the expense of farming activity and landscapes; the rural depopulation translates into a gradual abandonment the land.

- So-called linear forests, consisting of trees spaced to form hedgerows (acacia, ash), groves, orchards (apple, pear, mulberry, olive), river borders (poplar, alder) have receded dramatically, due to regrouping of lands, changes in farming activities or techniques, and wide-spread urbanization. An estimated 100 million trees and 500 000 linear km of hedgerows have disappeared over the last 30 years. Since the beginning of the last century, 3 million hectares have been transformed, and only 1.6 million hectares of linear forest remain.

- Resinous trees, which represented only one-quarter of the woodland area at the beginning of the last century, have gained ground and now represent about one-third. Pine, fir and spruce grow more rapidly than deciduous trees and are therefore more profitable in the short term. However, resinous trees increase the acidity of the soil and weaken the overall composition of the forest on account of their relative vulnerability to disease and pollution.

A less visible and therefore little known phenomenon, whose import is only beginning to be appreciated, concerns the erosion of the biological diversity of the soils. This diversity plays a fundamental role in primary production and recycling organic matter. Almost all over the world, the functioning, structure and properties of the soil are being modified by human activities: agriculture, pollution, and artificialization due to urbanization. This leads to:

- increased soil erosion due partly to deforestation, over-grazing, and intensive farming. The global soil loss is estimated at 5 to 7 million hectares per year;

- salination as a result of irrigation, especially in arid or semiarid regions;

- compaction, which follows from the use of unsuitable or too heavy farming equipment, from too much trampling by animals, poor irrigation practice, etc.;

- high pollution with toxic organic compounds or heavy metals (Cu, As, etc.) as a result of treating crops with chemical fungicides, pesticides, etc.

Almost everywhere, the soil is suffering from impoverished biological diversity and activity, as well as reduced organic matter content. This is an important aspect of the phenomenon of *desertification*, which refers to soil degradation in arid and semiarid zones as a result of climate variations and human activities. On the human scale, desertification manifests itself as a reduction in vegetation cover as well as a loss or destruction of the biological potential of soils and their capacity for supporting the populations that live there.

6.3.3 Biological invasions and the introduction of species

Human migrations have been occurring on a large scale for several tens of thousands of years. Whether or not they have involved the colonisation of new environments, they have unleashed processes of species transfer within continents, between continents and nearby islands, and between the continents themselves. When Neolithic man moved to new territory, he not only took along his domestic species but also introduced a number of accompanying wild species of plants and animals. Before the arrival of humans between 5 and 6 thousand years ago, the Mediterranean islands were populated by an endemic fauna inherited from the Tertiary era. It included a dwarf elephant (*Elefas falconeri*) in Sicily and a 'rabbit' (*Prolagus sardus*) in Corsica. Over the brief period of several thousand years, this entire fauna disappeared, supplanted by an exclusively modern fauna. Was this new fauna introduced deliberately or not? The question remains open.

What is a biological invasion?

The voluntary or accidental migration of individuals from one ecosystem to another is a relatively frequent phenomenon. Often, the immigrants encounter such unfavourable conditions that they make only a brief appearance in the new environment. But in other cases, the species finds all that it requires, appropriating resources used by indigenous species, and is able to propagate to the point where the latter are partially or entirely eliminated. Such immigrants may profoundly modify equilibriums established over thousands of years as well as the functioning of colonised ecosystems.

A species is called invasive if it colonizes and proliferates in an ecosystem where it was previously not present. Generally speaking, such colonizations may result from human actions or ensue from chance or accidental displacements associated with the dynamics of natural populations.

Species are considered *indigenous*, *native*, or *autochthonous* if they originated in the environment under consideration. They are described as *introduced*, *exotic*, *exogenous*, or *allochthonous* if they spread from another region of origin. When a new population is adapted to an environment, i.e. lives and reproduces there, this process is called *naturalization*.

6.3.3.1 Deliberate introductions

Many deliberate introductions have been motivated by a concern for increasing the local potential of species useful to humans. In most parts of the world, for example, the food demand is covered by plant and animal species originally deriving from other continents.

The 'discovery' of America revealed an ancient and relatively advanced form of agriculture based on three main plants: corn, cassava and the potato. Native American Indian farming was nonetheless diversified, and the Europeans took advantage of it to export about 20 South American plants out into the world: corn, cassava, potato, tomato, pumpkin, tobacco, strawberry, chilli pepper, bean, etc. Europe later also benefited from plants originally deriving from North America, such as the Jerusalem artichoke and sunflower. Turkeys and barbary ducks also profited by cultivation in Europe.

These transfers are behind the greatest revolution in the history of food production. Corn and potatoes played an important role in the agricultural dynamics of Europe and the implementation of a 'new agriculture' at the beginning of the 19th century. International transfers were also to play a role in the industrial revolution (cotton, rubber) and enriched the stock of medical products (quinine). There is a tendency to see Europe as the principal beneficiary of the discovery of South America and its domestic species. But South American species were also transferred to Africa and Asia: rubber, cotton, sisal and cocoa trees, as well as crops contributing towards reducing the risk of famine or food shortages: corn, cassava, the sweet potato, the peanut, etc. On the other hand, the American continent benefited greatly from the transfer of the horse, cattle, swine and fowl as well as wheat from Europe. Coffee and yams were imported from Africa, and rice, sugar cane, soy bean, banana, citrus and coconut trees from Asia, etc. So the New World did not emerge from these exchanges empty-handed.

Thus, the international transfer of plants and animals for the improvement of food production has been going on for several centuries. The process is still continuing, as illustrated, for example by the introduction of the kiwi fruit only a few decades ago. The world has gradually become a vast supermarket for biodiversity. If a species appears to be of any interest whatsoever, there will be an attempt to acclimatize it and turn it to profit. But there is a problem! The very biological characteristics valued in these species (rapid growth, adaptability, easy reproduction) are the same properties that make them good candidates for biological invasions.

Some 277 fish species have been introduced to Europe. Table 6.2 indicates that almost one-third of these introductions took place in the 1960s and 1970s. In France, scientists have counted 27 introduced species of fish, as compared with 31 in Great Britain, 19 in Germany, 43 in Italy and 20 in Belgium. Over one-third of Europe's current fish fauna is composed of introduced species.

6.3.3.2   Species that 'escape'

The introduction of species for aquaculture is practised almost all over the world. Many of the species introduced into captivity eventually spread out into the natural world. Thus, in December 1999, catastrophic floods in the Bordeaux region damaged pisciculture basins containing Siberian sturgeon (*Acipenser baeri*). Thousands of individuals of this species have dispersed throughout the Garonne basin, where its cousin, the *Acipenser sturio*, is in danger of extinction. It is anticipated that the two populations will cross-breed. This example is far from anecdotal;

Table 6.2 Number of fish species introduced to Europe by decade for some large taxonomic groups (after Cowx, 1997)

Decade	Salmonids	Coregonids	Cyprinids	Centrarchids	Cichlids	Others	Total
<1800			8			2	
1800–1845			1	4		3	8
1850							0
1860	2	1				1	4
1870	3		1	4		3	11
1880	11	6	2	6		5	30
1890	23		2	4		1	30
1900	3	1	1	4		3	12
1910	1			2		4	7
1920	3		1	2	2	4	11
1930	7	1	1	1		2	12
1940	4	1	3		1	3	12
1950	7	3	3	3	1	2	19
1960	12	1	27		2	9	51
1970	13	3	10	1	2	14	41
>1980	2		8		1	6	17
Total	91	17	68	31	8	62	227

rather, it represents the rule. Dozens of aquatic species have been 'accidentally' introduced in this way almost all over the world. Nowadays, it is actually possible to spot a flock of sacred ibis (*Threskiornis aethiopicus*) flying over the edge of the Gulf of Morbihan or in the marshes of the Guérande. Originally from tropical North Africa and well known to the Egyptians, who used to mummify them, these waders escaped from a zoological park and are now 'naturalized' inhabitants of France.

6.3.3.3 Chance or accidental introductions . . . ecological roulette

The development of international trade is the probable cause for the current, unprecedented circulation of thousands of plant and animal species as well as microbes around the world. Trade liberalization has multiplied the existing trade routes around the world, a situation that aggravates the risk of invasions.

As of the 1880s, the use of water as ballast in cargo ships unintentionally implemented a vast network of exchanges between the flora and fauna of biogeographic regions that had previously remained isolated. Cargo ships transported huge quantities of seawater or brackish water,

---

Miconia in Polynesia

One of the most disastrous cases of a plant invading an insular ecosystem is that of *Miconia calvenscens* in Polynesia. Growing to a size of up to 15 m, this tree is indigenous to the tropical rain forests of central America, where it colonizes small clearings in the forest. Introduced to Tahiti in 1937 as an ornamental plant valued for its superb dark-green leaves with purple backs, this tree has spread over the whole island in less than 50 years, supplanting the indigenous forests and forming dense, monotypic settlements. It currently covers 70% of the island, proliferating at distances between 10 and 1300 m in various different habitats. Its rapid growth (up to 1.5 m per year), early maturation (4 to 5 years), autopollination, prolific fruit and seeds, and the active dispersal of its plump bulbs by introduced frugivorous birds are factors that combine to make this species a particularly aggressive colonizer in competition with the indigenous insular species.

This plant has modified the structure and composition of the indigenous vegetation besides affecting other ecosystem processes (the distribution of light and water, the cycle of soil nutrients, etc.) *Myconia* has spread to neighbouring islands, where information campaigns and intensive combative efforts have helped to halt its expansion. However, it is still continuing to spread to other islands, where *Myconia* seeds are introduced accidentally in the soil of potted plants deriving from Tahiti.

---

which they drew from one port and released in another. This water contained many planktonic organisms and planktonic stages of benthic organisms. It has been one of the most significant mechanisms of transoceanic dispersal for aquatic organisms.

Sampling the ballast of 159 cargo ships arriving in the Coos Bay (Oregon) from Japan, scientists found 367 identifiable species belonging to a wide array of marine groups. All the trophic groups were represented. Not all of these species become naturalized, but depending on the circumstances, some of them became established in the new environment. In this way, many bays, estuaries, lagoons and lakes are experiencing a continuous influx of species whose adaptive abilities, roles and ecological impacts are not always predictable. This is what some people call 'ecological roulette'.

The example of the Great Lakes of North America is also symptomatic. Over the last century, the number of exotic species in the lakes has grown considerably. About 75% of the new varieties registered since 1970 are believed to have arrived in the ballast water of trans-oceanic ships coming from Eurasia. What is more, most of these species are native to the Ponto-Caspian region (Sea of Azov, Caspian and Black Seas)

6.3.4 Overexploitation

Scientific literature abounds with descriptions and denunciations of the effects of the overexploitation of living resources and the resulting extinction of species. It is true that humans have pursued and destroyed many species, or at least endangered their existence. Special mention might also be made of the European bison and many cetacean species that have almost disappeared, or the American passenger pigeon or the dodo of the island of Mauritius, which are both extinct. The list of endangered species is long, including tigers, pandas, rhinoceroses and a number of marine mammals. In some cases, they are killed to satisfy food demand or local needs. In other cases, the culprit is untimely hunting or hunting for profit. Thus, the fur industry has led to the destruction of a number of wild species, including the great predators (panther, ocelot, wolf, fox, etc.). In a different, less well-known domain, collectors (shells, insects, orchids, etc.) do trade in rare species and endanger certain endemic populations.

The pressures exerted upon the oceanic ecosystems and their resources are rising. Of the 200 most important fish stocks, representing 77 per cent of the total catch, 35 per cent are currently over-fished. The marine benthic fish of the North Atlantic (cod, haddock, hake, etc.) have suffered considerably due to over-intensive fishing over decades, to the point where certain fisheries are in danger of disappearing. Moreover, the trawlers which are used to catch them destroy the ocean floors, in addition to capturing large quantities of fish with no commercial value.

With the destruction of coastal ecosystems and increased coastal pollution, the situation has become critical in coastal regions. Around 50 per cent of the mangrove ecosystems, for example, have been transformed or destroyed. Corals are particularly affected by the intensive exploitation of their living resources and the modification of coastal

The Emperor penguin

The Emperor penguin (*Pinguinus impennis*) is a seabird, an excellent swimmer but awkward on land. Hundreds of thousands of individuals used to live in the North Atlantic region, where they were hunted for their feathers in the 18th and 19th centuries. The last specimen was killed in Iceland in 1844.

For a long time, it was believed that this bird, adapted to the cold, had abandoned the coast of Europe for the islands of the North Atlantic as the climate grew warmer. But recent findings have revealed that the Emperor penguin was actually well adapted to temperate climates, since it was present in the Netherlands around 2000 BC. It now seems clear that this animal, which was unable to fly and therefore highly vulnerable, was valued by prehistoric man for its flesh. It appears to have been chased away from its area of origin and have taken refuge on the North Atlantic islands, where it was later hunted to the point of extinction.

environments: changes in land use leading to soil erosion and decreased water transparency, fertilizer run-off and domestic wastes, pollution by pesticides and petroleum products, etc. Another factor is tourism, which degrades coral populations with inopportune harvesting and increases the outbreak of diseases. Not to be disregarded are the consequences of El Niño, causing a rise in temperature over recent years that is responsible for the phenomenon known as coral bleaching (expulsion of the zooxanthellae that live in symbiosis with corals).

Another domain that has received considerable publicity concerns the destruction of the tropical forests. The loss of thousands of hectares every year is the result of two quite specific activities:

- the exploitation of timber for export. Too frequently, foresters use methods that involve less work but are extremely destructive for the ecosystem;

- the search for farmland and pastures, leading to the destruction of vast forest surfaces by fire. This is a frequent practice in many parts of the world.

6.3.5 Some examples illustrating the combined effect of human activities upon the dynamics of biological diversity

Very often, the erosion of biological diversity results from the combined effect of multiple factors acting in synergy. Two examples follow by way of illustration.

6.3.5.1 The extinction of cichlid fish in Lake Victoria

One of the lamented consequences of introductions is the elimination of indigenous species by the introduced species. Many studies amount to veritable necrologies of extinct species, although the exact role of the introduced species is not always known. In reality, extinctions often result from the simultaneous action of different factors that may have a synergetic effect. One well-known example is that of Lake Victoria in East Africa. Toward the end of the 1950s, a large predator fish called *Lates niloticus* or Nile perch was introduced. Present in the large river basins from Senegal to the Nile, as well as in Zaire, this species is prized for consumption and sports fishing. At the time, Lake Victoria was home to several hundred fish species of the cichlid family, all participating in a highly complex trophic network. During the 1980s, the development of *Lates* populations was accompanied by the near extinction of dozens of cichlid species that fed off phytoplankton and zooplankton. Subsequently, many other species also became scarce, whether or not they belonged to the cichlid family. On the level of ecosystem functioning and trophic networks, many autochthonous fish were replaced by only two indigenous species: the detrivorous shrimp *Caridina nilotica*, and a fish that feeds on zooplankton, the cyprinid *Rastrineobola argentea*. These two species have developed abundantly, and since the disappearance of the cichlids, they serve as food for *Lates* larvae and juveniles. Thus, the food relationships among species in the ecosystem of Lake Victoria have been drastically simplified, with cannibalism playing an important role: adult *Lates* eat their young. This observation led the scientific community to stigmatize the introduction of *Lates niloticus* as having caused an ecological disaster. Scientists lamented the loss of a veritable laboratory for evolutionary studies, since the hundreds of endemic cichlid species had provided an excellent model for speciation in progress. Meanwhile, other studies focused on the changing living conditions in Lake Victoria as a consequence of developments in farming and urbanization upstream from the lake. Fertilizer wash-off triggered a process of eutrophication, which manifested itself in a significant increase in algae development, reduced water transparency, and deoxygenation of deep waters – all of which are factors equally prone to disturb cichlid reproduction dramatically. In the 1970s, the introduction of new fishing techniques, such as bottom trawling, also weakened the cichlid populations, most of which reproduce at the bottom of the lake. And finally, in recent years, Lake Victoria has been invaded by the water hyacinth, which also modifies the ecosystem. Thus, the consequences of the introduction of a large predator like *Lates* upon lacustrine fauna must be regarded in the more general context of environmental changes due to human activities. Without denying the effect of the predator, the fact that the fish populations of Lake Victoria were already weakened by other disturbances is probably an equally important factor.

6.3.5.2   The Mediterranean Sea

Observations over recent decades provide evidence for changes to the biological diversity of the Mediterranean Sea as a result of several related factors.

- A rise in water temperature. This tendency is especially well documented for the Ligurian Sea, which is, in principle, the coolest part of the Mediterranean. The presence of warm-water marine species in this zone has increased dramatically since 1985.

- The opening of the 163-km long Suez Canal in 1869 enabled considerable movement and exchanges between the Red Sea and the Mediterranean, two biogeographic provinces that had been separated for almost 20 million years. Nearly 300 species from the Red Sea and the Indian Ocean penetrated into the eastern Mediterranean by this route and settled there. An estimated 60 per cent of the species have been introduced since the 1970s. In the present biogeographic context, this constitutes the most spectacular biological invasion in a marine environment. The flux has been almost entirely one-directional; very few species have migrated from the Mediterranean to the Red Sea. Today, 'Lessepsian migrants' represent around 4 per cent of the species diversity in the Mediterranean and 10 per cent of the diversity on the Levantine Coast. Most of the species involved in these exchanges are benthic and demersal species, in particular algae, molluscs, crustaceans and fish.

- 'Accidental introductions'. A word should be said about one of the most publicised biological invasions in France: the expansion of the tropical alga *Caulerpa taxifolia* along the Mediterranean coasts. Accidentally introduced to the coast of Monaco, this alga from the Pacific quickly adapted to its new habitat. It has developed at a pace unknown even in its place of origin and has overrun the French and Italian coasts.

7 The Dynamics of Biological Diversity and Implications for Human Health

Human beings are a product of evolution like other living species and an integral part of biological diversity. They are considered a 'success story', because the success of the human species, measured in terms of its capacity for adaptation, socialization, and technical proficiency, is without comparison. Nevertheless, the human species remains subject to the same biological constraints as other species. The coevolution of human beings and other living organisms has forged a system of complex relationships, most notably host–parasite relationships. Environmental changes impact human behaviour and the established relationships of humans with other species.

7.1 The Complexity of Host–Parasite Relationships

Like other mammals, humans are host to a number of parasites. Often, the biological cycle of these parasites involves at least two hosts: one final host, and one or more intermediary hosts in which they accomplish a part of their life cycle. These intermediary hosts transmit the parasite to humans: mosquitoes are the vector for malaria, blackflies for onchocerciasis (river blindness), cats for toxoplasmosis, etc.

To study the epidemiology of parasitic diseases, a thorough knowledge of taxonomy is required:

Biodiversity   Christian Lévêque and Jean-Claude Mounolou
© 2004 John Wiley & Sons, Ltd   ISBN 0 470 84956 8 (Hbk)   ISBN 0 470 84957 6 (pbk)

- to identify the morphology of disease vectors and reservoirs, which may be either vertebrates or invertebrates;

- to establish the cytogenetic or genetic identities of the different strains of vectors or pathogens involved. It is becoming increasingly evident that small genetic variations can have a serious impact upon parasitic species' capacity for transmission or degree of pathogenicity.

7.1.1 Human onchocerciasis

Research on human filaria (*Onchocerca volvulus*) in western Africa has demonstrated the importance of a precise identification of parasitic vectors and infective strains in order to understand the different dynamics of disease transmission over zones with extremely varied climatic and phytogeographical characteristics.

The vector in question is the blackfly, a dipter first identified as a single species: *Simulium damnosum*. Later, it became apparent that the species exhibited biological differences depending upon the bioclimatic region involved. Cytogenetic research revealed the existence of a set of twin species that were morphologically very close (and thus difficult to distinguish) but differentiated by their chromosome structures. In western Africa, there are three subcomplexes (*S. damnosum*, *S. sanctipauli* and *S. squamosum*), each comprising several species. A significant epidemiological discovery showed that the different West African species of the *S. damnosum* complex (nine in all) do not occupy the same phytogeographical zones. Some are typical of savanna zones, while others are found primarily in forest zones. It has been demonstrated that these different species do not have the same capacity for transmitting onchocerciasis: forest varieties are more active vectors of the parasite than savanna varieties.

On the other hand, there are also several different species of parasites, and black flies can also act as the bearers of microfilariae that are morphologically close to the human parasite (*O. volvulus*) but infect other animals. In order to implement measures for controlling the disease, it was important to know which species was involved. In view of the difficulty of identifying the strains of the parasites on the basis of their morphologies, DNA probes were developed to distinguish *O. volvulus* from the other species.

Finally, for the filaria *O. volvulus* itself, differences in degree of pathogenicity were established between forest strains and savanna strains. Here too, the use of DNA probes made it possible to differentiate among them and thus to refine strategies for combating the disease.

This example illustrates the importance of applying highly sophisticated tools of molecular genetics (molecular markers) towards understanding the complex ways in which pathogenic species are transmitted and avoiding erroneous epidemiological conclusions. It is necessary to know the different strains of vectors in order to evaluate their capacity for transmission; human microfilariae must be clearly distinguished from animal microfilariae; the pathogenicity of different strains of microfiliaria is key to understanding the dynamics of transmission in forests or savannas, etc.

7.1.2 Malaria

Malaria is one of the most lethal diseases of the planet, responsible for over 2 million deaths each year. The causal agent is a protozoan, *Plasmodium*, transmitted by mosquitoes of the *Anopheles* species. Several species of *Plasmodium* are pathogenic for humans. In the 1940s, effective and inexpensive insecticides (DDT, for example) and antimalaria drugs were developed, and it was anticipated that this endemism would soon be checked. As yet, however, there is no end in sight.

One of the main difficulties in the battle against malaria resides in the variety and complexity of its transmission system. In Africa, five species of *A nopheles* act as effective vectors: *A. gambiae*, *A. arabiensis*, *A. funestus*, *A. nili* and *A. moucheti*. But there are also another eight or nine vector species of local or secondary importance. As a result, a number of different agents may transmit malaria in the same place – sometimes simultaneously, sometimes during different seasons. Moreover, each type is characterized by great intraspecific variability. *A. funestus* belongs to a group of closely related species that are difficult to distinguish and for the most part considered to be zoophilous. *A. funestus* itself is extremely heterogeneous: some populations are basically anthropophilous, while others are partly zoophilous, and their ability to transmit *Plasmodium* also varies.

At first, a single species of *A. gambiae*, the principal vector of malaria, was identified; however, it is now known that six twin species are involved. There are also cryptic species and 'cytological forms' whose frequency varies with ecological conditions and seasons. The complexity of the genetic systems of *A. gambiae* is extraordinary: from intraspecific polymorphism to new taxa emerging through genetic exchanges between related species (introgression). This complexity is probably due to the species' capacity to adapt rapidly to evolving environments, in particular to man-made habitats.

Finally, some species that are normally considered to be secondary may become powerful vectors in certain environments. The difficulties encountered in controlling malaria also stem from the fact that the species or 'forms' do not share the same biological and ecological characteristics, meaning that the methods of combat must be diversified.

But the complexity of the problem also resides in the genetic variability of the parasite *Plasmodium falciparum*, which is capable of undergoing extreme antigenic variations in order to slip past the immune response of the host. In turn, humans defend themselves by developing a partial immunity that grows stronger as contaminations accumulate over the course of life.

7.2   Emerging Pathologies

For a long time, viral and bacterial infections were the main causes for human mortality. The influenza epidemic that raged in 1918/1919 (Spanish flu) killed between 20 and 40 million people – more than the First World War. This was probably one of the most severe natural catastrophes ever to hit humankind. It was preceded, in the early Christian era, by three great epidemics of bubonic plague, an animal disease that is transmitted to humans through the bites of fleas from infected rodents.

With improved hygiene and vaccinations, this type of mortality receded considerably, to the point where there was hope of eventually resolving the problem of infectious diseases. During the last decade, however, the trend has reversed. Old diseases have reappeared and spread to new zones, sometimes due to the pathogens' acquired resistance to drug medication. This is the case for malaria, tuberculosis, yellow fever and cholera. Linked to expanded urbanization, the appearance of forms resistant to therapeutic treatments has favoured the return of tuberculosis, which killed 3 million people in 1995.

In addition to old infectious diseases that remain a significant cause for mortalities, new, hitherto unidentified diseases are appearing around the world and claiming large numbers of victims. These new diseases are known as emerging diseases. Around 30 have been identified since the early 1970s. One such disease is hepatitis C, caused by a virus discovered in 1989, which claims tens of thousands of lives each year. Other examples are human immunodeficiency virus (HIV), Ebola virus, Lyme disease, etc.

Many emerging diseases are caused by pathogens that have been present in the environment for a long time but have only recently come into contact with humans via another species and in the wake of environmental changes increasing the chances of contact. One example of an emerging disease is the new variant of Creutzfeldt–Jakob disease, first described in the United Kingdom in 1996. Its agent is held to be identical to that of bovine spongiform encephalopathy (BSE), which appeared during the 1980s and hit thousands of cattle in Europe.

What are the reasons behind this perceived development? First, there has been considerable progress in the analysis of the micro-organisms responsible for infectious diseases. While these have been known for a long time, until recently, cultivation techniques were not sufficiently advanced for the causal pathogenic agents to be isolated. Second, the appearance of certain diseases is linked to disruptions in our ecosystems and behaviour. The example of Lyme disease, which caused an epidemic in the United States, is a case in point. The causal pathogen is a spirochete called *Borrelia burgdorferi*. The advance of the disease is linked to the proliferation of deer in the northeast of the United States, where they are no longer hunted and many houses have large, unfenced gardens in the proximity of woods. Ticks bearing the spirochaetes are imported by cervids into inhabited zones, where they bite humans. This leads to a large increase in this infectious disease in exposed regions. Altogether, ticks are at the root of a number of new diseases, such as the various forms of rickettsiosis in Japan and Africa.

Numerous instances of emerging diseases have also been noted among wild animals. This has a number of different causes:

- transmission from domestic animals to wild species living in the vicinity. One example is the transmission of the canine morbillivirus (Carré's disease) to the lion, causing high mortalities in Africa's Serengeti National Park in 1991;

- the increasingly observed transmission of pathogens from one wild species to another. Massive mortalities have recently been registered among marine animals such as mammals and corals as a result of more frequent epidemics and the appearance of new diseases. The origin of many of these new diseases lies in the transmission of a known pathogen to a new host, rather than the appearance of new pathogens. Climatic changes and human activities have accelerated the migration of species around the world. In the process, hosts have come into contact with pathogens to which they were not previously exposed. This would explain the explosion of new diseases in different species.

7.3 Environmental Change, Biological Diversity and Human Health

Environmental changes account for the appearance or development of numerous diseases affecting human populations. Extended irrigation and water reservoirs both favour the propagation of vectors such as mosquitoes and molluscs. The growth in world population and occupation of new territories increase the probability of contact between humans and the vector species of pathogenic organisms. International trade and manifold intercontinental exchanges also lead to the dissemination of pathogens or their vectors. Other factors involve changes in lifestyle and the concentration of people in cities, favouring transmission from person to person. In place of an exhaustive list of consequences that changes to the environment and biological diversity hold for human health, the examples given below will serve to illustrate the diversity and scale of such phenomena.

7.3.1 Intercontinental exchanges

One of the major risks of international exchanges is the possibility that viruses, bacteria, fungi, protozoa and parasites, as well as their vectors or potential reservoirs, may be introduced into a receptive environment. There are many examples of mass mortalities in human populations or wild species following the introduction of new pathogenic agents

Human beings may themselves act as vectors for infectious agents. Thus, several million people in the New World died when they came into contact with the infectious diseases borne by the conquistadors. The smallpox imported into Mexico by the Spanish army in 1520 killed 3.5 million people – half of the indigenous population, in just 2 years. Still in our day, Amazon tribes may be decimated by exposure to new infectious agents such as measles.

The situation is more complex when the cycle of the infectious agent includes an intermediary host and/or vector. In the 17th century, boats transporting slaves from Africa to the Caribbean carried supplies of fresh water infested by the mosquito *Aedes aegypti*, the vector for yellow fever and dengue haemorrhagic fever. The slaves provided the reservoir for the virus. Until today, *A. aegypti* has withstood numerous eradication campaigns and still constitutes the major vector for urban yellow fever.

The mosquito *A. albopictus* was introduced from Japan to the United States in 1985 when used tyres were shipped to American factories for reprocessing rubber. Their larvae survived in the inner tyre tubes where some rainwater had collected. Spreading rapidly, the mosquito can carry dengue fever as well as the arboviruses responsible for encephalitis. This growing species continues to colonize new territories. Meanwhile, *A. albopictus* has also arrived in Brazil, Nigeria, Italy and New Zealand.

Malaria presents a rather similar case. *Anopheles gambiae* was introduced to the northeast of Brazil in 1930 by a ship from western Africa. Over subsequent years, the disease spread to the point where it became necessary to implement an eradication programme on an unprecedented scale to eliminate the mosquito – with success – the last *Anopheles* in the region was caught in 1940.

There are also numerous examples among domestic animals for the transfer of parasites from one continent to another and the transmission of pathogens to related species. Toward the end of the last century, European crayfish were widely decimated by a hitherto unknown epidemic, the 'crayfish epidemic', a rapidly propagating epizooty deriving, as was later discovered, from a fungus and probably transported from the Mississippi to Italy in the fresh water supply of commercial cargo ships. This disease still erupts sporadically in Europe. During the 1970s, the flat oyster *Ostrea edulis*, which is central to bivalve production in France, was decimated by a parasite, *Bonamia ostreae*, which was probably introduced from the west coast of the United States in shipments of oyster spats.

7.3.2 New technologies related to lifestyle

The introduction of new technologies also causes humans to enter into relationships with micro-organisms with which they had no previous contact. A famous case is Legionnaire's disease. Following a meeting of members of the American Legion in 1976, a number of veterans succumbed to a form of pneumonia accompanied by high fever. The malefactor was isolated and identified: a stick-shaped bacterium that was given the name *Legionella pneumophila*. Since then, many cases have been registered almost everywhere in the world. However, it is difficult to diagnose Legionnaire's disease, because the genus *Legionella* comprises several dozen species, with 15 serological groups identified so far. In 85 per cent of the human infections, the implicated variety is *L. pneumophila* of serogroup 1, but 5 per cent of the cases involve one of the other 14 serogroups. The remaining 10% of these infections are caused by species other than *L. pneumophila*. The most frequent sources of contamination are hot water reservoirs (showers, jacuzzis, decorative water fountains), and air conditioning systems; transmission is through inhalation of bacteria contained in the water vapour clouds emitted by such devices. The number of cases declared in France and in the world is clearly on the increase (582 cases in the year 2000 in France).

In the food domain, changes in behaviour have also contributed towards the emergence of new pathologies or the return of rare diseases known from former times. Thus, the incidence of salmonellosis epidemics has risen over the last 20 years in Europe and North America, where the *enteritidis* serotype of *Salmonella* has become the predominant strain. This infection is largely linked to the consumption of contaminated eggs. Beginning with a unique clone discovered in 1982, the 'hamburger bacteria' *Escherichia coli* O157 underwent a spectacular development, causing hundreds of thousands of infections around the world. As for cases of listeriosis, they usually follow from the consumption of cheese or processed meats (pig's tongue and rillettes, for example) that have undergone prolonged cold storage, since the pathogenic agent *Listeria monocytogenes* is actually capable of multiplying at low temperatures. These examples illustrate the consequences of changes in lifestyle relating to the consumption of industrialized and fast food.

7.3.3   The eutrophication of water and the proliferation of toxic algae

All the coasts of Europe, as well as other shorelines on the planet, experience a periodic proliferation of microscopic algae, sometimes described as 'red tides.' The organisms responsible for this phenomenon are mostly dinoflagellates, of which approximately 40 toxic species have been identified in the world. They produce toxins that can cause mortalities in marine fauna and occasionally fatal intoxications in humans. Most often, contamination occurs through the consumption of shellfish that have themselves ingested toxic algae. The symptoms are abdominal pains, diarrhoea and vomiting. The toxic substances, some of which are among the most lethal known, can lead to paralysis of the nervous system and consequent death by asphyxiation.

While these algae have been known to exist for a long time, both the frequency and scale of their appearance have increased considerably over recent years. Several thousand cases of intoxication are registered annually in France. One possible explanation points to the ecological disequilibrium caused by human activities. Soil fertilizers that are washed away by rains and borne towards the coastal regions by rivers, together with domestic wastes rich in nutrients, enrich seawaters and favour algal proliferation. In freshwater environments, cyanobacteria develop in a manner similar to dinoflagellates in eutrophicated environments, likewise giving off toxins that are dangerous for humans.

Ciguatera or 'fish poisoning' is caused by marine neurotoxins found in various zones in the Pacific and Indian Oceans. For a long time, scientist believed that these toxins were produced by several varieties of marine fish; but actually, the responsible organism is the dinoflagellate *Gambierdiscus toxicus*, which lives as an epiphyte of large algae or corals and is consumed by fish.

7.3.4 Allergies

Allergies are often pathological expressions of anaphylaxis. They are perceived in terms of an immediate hypersensitivity, because the reactive effect follows within minutes after the stimulation. The immunological process involves induction by an antigen (or allergen) and synthesis of a particular type of antibody, namely immunoglobulin E (IgE). The extrinsic factors inciting the production of IgE in humans are usually found in the environment. They are harmless for most individuals; i.e. their impact upon healthy subjects is no different from that of ordinary antigens, while for allergic patients, they act as allergens.

Airborne allergens that cause respiratory problems derive from various sources:

- *Pollens* are responsible for 50 per cent of the allergy cases in France, affecting about 4 per cent of the population. The main sources are pollens of grasses and trees. The introduction of plants deriving from other continents can increase the risk of allergic reactions.

Ambrosia (*Ambrosia artemissifolia*), also known as ragweed, belongs to the sunflower family. It originated in America and was introduced to Europe from Argentina toward the end of the 19th century, together with the red clover and potato plants. This pioneer plant is highly opportunistic and thrives on bare ground, such as fallow lands and at roadsides, but also on cultivated fields and land with sparse vegetation. Its pollen, which is very fine and especially abundant, acts as a powerful allergen. A single stalk can emit 2.5 billion grains of pollen, which the wind can transport as far as 100 km from their point of origin. Already one to three grains of pollen per cubic metre of air are enough to cause problems for sensitive persons. At the height of pollination (from August to October), up to 100 grains of pollen are registered per cubic metre in the Rhône-Alpes region of France. Since nearly 10% of the population is allergic, this creates a serious public health problem. It seems that the recent proliferation of ambrosia in this part of France is a result of the freeze on agricultural cultivation imposed by the EU Commission in Brussels in the early 1990s. Eradication campaigns have met with only partial success.

- *Mites* of the genus *Dermatophagoides* are found in household dust and are responsible for most cases of allergically induced asthma. Their true ecological niche is the mattress, but they also thrive in rugs and carpets, as well as in stored foodstuffs. Thus, modern life styles have also contributed towards their proliferation.

7.4   The Growing Virulence of Viral Diseases

Infectious diseases are manifestations of the relationship of humans with their environment and the reproductive strategies of pathogenic micro-organisms that find fertile grounds in humans. The growth of the human population and its increased contact with other animals create conditions that favour the selection of humans as the preferred host for certain micro-organisms. The emergence of new viruses, most notably those causing haemorrhagic fevers (such as the Ebola and dengue viruses) and HIV, is of growing concern to health services (see Table 7.1).

How can a virus emerge and generate a hitherto unidentified disease that afflicts humans? On the one hand, given their rapid life cycles, viruses are able to adapt quickly to environmental changes, as compared with humans and other animals with longer life cycles. Thus, the emergence of a virus may result from the *de novo* evolution of a new viral variant in the wake of mutations or recombinations between existing viruses, possibly engendering more virulent strains. The influenza A virus, for example, is constantly evolving, and new strains propagate in human populations. Since they are new, hardly anyone is immune to them. Vaccination only protects against the familiar strains that were used to develop the vaccines. Encountering no resistance, new strains spread rapidly throughout the world and can cause a great number of deaths.

On the other hand, certain viruses that have existed for a long time among particular animal species may suddenly appear in the human species as the result of environmental change. Thus, creating cultivated zones by deforestation facilitates the proliferation of rodents, which frequently serve as animal reservoirs for viruses; the construction of dams favours the rapid propagation of mosquitoes, which act as vectors for

Table 7.1 Some examples of emerging diseases

Virus	Symptoms	Distribution	Natural hosts	Causes of emergence
Yellow fever	Fever	Africa, South America	Mosquitoes, monkeys	Urbanization, dams
Dengue fever	Haemorrhagic or non-haemorrhagic fever	Africa, Asia	Mosquitoes, humans/ monkeys	Urbanization, dams
Rift Valley fever	Haemorrhagic or non-haemorrhagic fever	Africa	Mosquitoes, ungulates	Dams, irrigation
Hantaan fever	Haemorrhagic fever	Asia, Europe, United States	Rodents	Contact with rodents
Lassa fever	Haemorrhagic fever	Africa	Rodents	Rodents
Ebola disease	Haemorrhagic fever	Africa	Unknown	Small mammals

many pathogens, etc. There are any number of factors that multiply the possibilities of contact between viruses and new hosts such as humans. Another important phenomenon deserves mention, namely the increasing transfer of these diseases and their vectors around the world.

For an emerging micro-organism to become a menace to public health, it must pass through two stages:

- *introduction of the virus into a new host, humans in this case.* This phase presupposes the existence of a reservoir in at least one non-domestic animal and brings into play mechanisms of transmission between the animals and humans, usually by means of animal vectors. The most significant animal reservoirs are rodents and arthropods (insects, ticks, etc.). Hundreds of viruses have been iden-tified in arthropods (arboviruses), of which at least a hundred can cause diseases in humans;

- *dissemination of micro-organisms in the populations of the new host*, which presupposes that adequate mechanisms of propagation are activated. The displacement of populations as a result of conflicts, urbanisation and travel (on the part of humans or other vectors) are all factors that favour the dissemination of viruses. HIV, the virus causing acquired immune deficiency syndrome (AIDS), was first dis-covered in Africa in 1981, whence it spread to the American continent and on to Europe and Asia. It is a known fact that the virus existed as early as 1970 and had already spread throughout Africa between 1970 and 1980. The identification of viruses related to the HIV in such varied animals as sheep, cats, horses and goats indicates that the HIV family goes back a long time. Various African monkeys are naturally infected with retroviruses that are akin to HIV but do not cause AIDS in these animals. One of the human viruses, HIV2, is very close to the virus of the Mangabey monkey of western Africa, and it is possible that humans may have been contaminated through bite wounds. As for HIV1, it could have been transmitted by chimpanzees, of which some are the bearers of a closely related virus. But it is not possible to establish precisely when these viruses managed to cross the species barrier. According to other hypotheses, humans have been contaminated for a long time, but the virus was not very widespread or virulent. The current epidemic could be the result of two simultan-eous developments favouring the diffusion of the disease: an evolution in the pathogenicity of human HIV, and changes in social behaviour.

New micro-organisms at the origin of human diseases continue to be detected. Some examples follow.

7.4.1 Morbilliviruses

The most familiar morbillivirus is the measles virus, which still kills at least 1 million people in the world each year, despite an effective vaccine. Other familiar viruses of this family cause Carré's disease in dogs and an analogous disease in cattle. However, over the last decade, other morbilliviruses have 'emerged', doubling the number of known viruses. A 'seal morbillivirus' was identified in animals that died in large numbers in 1988 on the coasts of northern Europe and were probably contaminated through farmed minks. A related virus was identified in the seals of Lake Baikal, which also died in large numbers. The virus is closely related to that of Carré's disease and is believed to have been transmitted to the seals by dogs living near the lake. Similarly, in the national park of Serengeti in Tanzania, a considerable number of lions died of a disease bearing a strong resemblance to Carré's disease. The big cats have become the new host of this canine virus, which has also caused an epizooty among spotted hyenas in the same region.

7.4.2 Haemorrhagic fever viruses

The most familiar of these viruses is the amaril virus, a Flavivirus responsible for yellow fever. This virus also infects monkeys, implying that yellow fever is a form of zoonosis that may have existed for thousands of years before being discovered in humans. The virus originated on the continent of Africa and probably reached the New World at the same time as its mosquito vector of the genus *Aedes*. Formerly restricted to savannas and forest rims, African epidemics, which are by far the most severe, are now penetrating the expanding urban settlements, which are providing new shelters for the mosquitoes.

Other varieties of Flavivirus have been known for a long time, e.g. dengue or 'tropical flu,' which is also transmitted by mosquitoes (*Aedes aegypti*). A relatively benign illness, dengue fever may be caused by any one of four types of viruses. The most serious, haemorrhagic form appeared in the early 1950s in southeast Asia and later spread to the Indian Ocean and Latin America.

The 'new viruses' of haemorrhagic fevers, some of which have been known for only a few years, belong to families other than the Flavivirus.

- *Rift Valley Fever* virus was originally a cause for epizootics. It started causing human deaths as of 1970 with the construction of the Aswan Dam, which favoured the rapid propagation of mosquitoes, whilst also creating human and animal concentrations favourable to contamination.

- *Hantaan virus* (Hantavirus) caused numerous deaths during the Korean War in the early 1950s. It was isolated in 1976 in the lungs of field mice, its primary reservoir in Korea. Thousands of people are infected each year, especially in China.

- *Ebola virus* was discovered in 1977. One of the most harmful pathogens for humans, it has caused several epidemics in Africa. It also afflicts monkeys. Lassa fever, endemic to West Africa, is a haemorrhagic disease similar to Ebola. The virus is present in rodents, and contamination occurs through contact with rodent excrements.

Emerging diseases, viruses and rodents

There are always many occasions for interaction between human and rodent populations, both in urban environments and in agricultural and forest environments. Rodents and humans have a great many disease agents and parasites in common. These are the 'natural reservoirs' of viruses that can be transmitted to humans. For example, the field mouse and vole are the habitual hosts of Hantavirus; in North America, strains of the Sin Nombre virus are borne by white-footed mice; in South America, several species of rodents serve as hosts to viruses of the Arenavirus family; and in western Africa, several species of rats (*Mastomys* sp.) appear to be similarly involved in the transmission of the virus for Lassa fever.

The risks inherent in human/rodent interactions increase with the density of rodents, the diversity of their species and their proximity to humans. The creation of new human habitats may favour the rapid proliferation of certain types of rodents, thereby increasing the risk of contact. Likewise, human behaviour may develop activities (deforestation, for example) that bring them closer to certain species of rodents with which they previously had little contact.

Through boat travel, Europeans also introduced the rat, *Rattus norvegicus*, into numerous regions of the world. This species transmits the bacteria of the plague, *Yersinia pestis*. North America experienced a progressive switch-over of the bacteria to autochthonous rodents, to the point where the reservoir is now permanent and rural. Although the plague is currently under control in most parts of the world, it is nevertheless quite possible that it may reappear at any moment. In Peru, for example, it caused several dozen deaths in the early 1990s.

Infectious diseases can be seen as the fruit of a dynamic coevolution. On the one hand, there is the evolution of micro-organisms that adapt to their hosts, thanks to their genomic flexibility and their extraordinary variability. On the other hand, the evolution of infected populations leads to the selection of more resistant individuals. Initial contact with an unknown infectious agent may unleash devastating epidemics, but with the selection of more resistant individuals, populations emerge that are capable of surviving the new infectious agent.

7.5   Adaptive Strategies of Pathogenic Agents and their Vectors

Humans have developed ways of combating pathogens. There were times when it was believed that, thanks to the progress of science, certain contagious diseases could be definitively eradicated. This hope was generally short-lived, since pathogens and their vectors have devised a whole series of adaptive strategies for surviving eradication campaigns.

7.5.1   Resistance to antibiotics

Antibiotics are natural chemical substances produced by certain micro-organisms (fungi, soil bacteria). In weak concentrations, they have the property of destroying or inhibiting the growth of other micro-organisms. The first antibiotic (the famous penicillin) was discovered in

1928 by Alexander Fleming and first used for therapeutic purposes in 1941. For several decades, antibiotics have made it possible to control the pathogenic bacteria responsible for major epidemics.

Early successes suggested that it would be possible to master the totality of infectious diseases. But because the plasticity of their genome enables bacteria to adapt and survive, the almost inevitable corollary of antibiotic therapies is the emergence of resistant bacteria. In some cases, a mutation occurs in a bacterial chromosome. In other cases, the bacteria acquire genetic information (plasmid) from other bacteria that are already resistant. The spread of resistance through the circulation of genes among bacteria is a frequent occurrence and explains the speed with which the phenomenon of resistance develops in the bacterial world.

The emergence of bacteria that are resistant to antibiotics is a worrying phenomenon of great concern to public health. When penicillin was first used in 1941, less than 1 per cent of the strains of *Staphyloccocus aureus* were resistant to this antibiotic; by 1994, 90 per cent of the strains of this same micro-organism had become resistant to penicillin, either through mutations that modified the targets of the antibiotic, or else by acquiring genetic information from other bacteria enabling them to destroy or neutralise the antibiotic. In the past few years, several species of bacteria have become resistant to all of the approximately 200 antibiotics developed over the last 60 years, and the problem will continue to grow.

Infections contracted in hospitals (known as nosocomial infections) present a serious problem for public health. Hospitals provide propitious conditions for developing resistance to antibiotics: intensive prescription of antibiotics, favouring the emergence of resistant bacteria, combined with a population concentration favouring the rapid dissemination of resistant strains. The higher the level of resistance, the more antibiotics are prescribed, leading to a fateful upward spiral of resistance. Without the discovery of new antibiotics, people risk finding themselves defenceless against serious infections.

7.5.2 Resistance to pesticides

Emerging resistance to the chemical products used to combat undesirable organisms (pathogens, disease vectors, agricultural pests) is becoming an increasingly acute phenomenon that affects all ramifications of life, from bacteria to eukaryotes of the highest evolutionary level. The resistance of insects reduces the effectiveness of insecticides, with serious repercussions for veterinary medicine (antibiotic treatment of livestock), agriculture (plant protection) and human health (treatment against the disease vectors and 'nuisances').

Resistance to insecticides is an expression of the capacity of organisms to adapt to new environmental conditions. The evolution of this capacity is dependent upon several factors:

- *mutations*, which result in new variants or alleles of existing genes, including the alleles responsible for resistance;

- *selection*, which sorts out the genes best adapted to the environment. In the presence of insecticides, resistant genes are selected and gradually become more frequent;

- *migration*, which enables these new genes to disperse beyond their geographic zone of origin.

Insecticides exert their toxic impact by fixating upon biological receptors – mostly proteins of the nervous system – and thereby disturbing their functioning. These molecules, the targets of the insecticides, can undergo point mutations enabling the attacked organisms to survive dosages of toxic substances that would normally be fatal.

Another way of reducing toxicity is by modifying the activity of detoxification enzymes, which are mostly esterases. In resistant individuals, there is an overproduction of these proteins that traps practically any pesticide molecule penetrating the organism. Among the biochemical and genetic processes involved, one of the best known is the amplification of the genes that code for proteins with esterasic activity. In the case of the mosquito *Culex pipiens* which inhabits the south of France, two principal mechanisms promote its resistance to organophosphate compounds, the insecticides most frequently used against this insect. First, enhanced detoxification occurs as a result of the overproduction of two enzymes, the esterases A and B. The overproduction of esterase B is caused by amplification of the gene coding this enzyme in the genome. This is followed by the emergence of resistant acetycholinesterases, proteins that affect the functioning of the central nervous system and that would normally be inhibited by organophosphate compounds.

Mosquitoes Fight back

Mosquitoes in West Africa are combated by spraying insecticides and using mosquito nets impregnated with pyrethrinoids. To some extent, these methods are countered by the mutated *KDR* (*knock down resistance*) gene that prevents the insecticide from taking hold in the organism of the mosquito. This form of resistance appeared in the wake of a massive treatment of the cotton fields of the Ivory Coast and spread over many regions of western Africa, with a prevalence sometimes exceeding 90%. The diffusion of *KDR* varies with regions and species of *Anopheles*.

7.6   Substances of Medical Interest and Biological Diversity

Biological diversity is an important source of natural substances with active components that are of interest to the pharmaceutical industry.

7.6.1   Traditional pharmacology

By default, populations with no recourse to chemical pharmacopoeia frequently resort to traditional pharmacopoeia. For a long time, plants were a major source of medicines. Opium, derived from the opium poppy, and its constituents (morphine and its derivatives) are the most familiar forms of ancient medicine, because they are effective against pain. Later discoveries were alkaloids and isolated active components of plants such as hemlock, cinchona (yielding quinine and its derivatives), and digitalin (an extract of digitalis), which counteracts certain cardiac deficiencies.

For economic reasons, around 80 per cent of human beings still have no access to modern medicine and treat themselves with traditional medicines, frequently derived from medicinal plants. Around the world, a total of approximately 20 000 plants are used in traditional medicine, whilst only 5000 have been scientifically investigated as potential sources of medical substances. This is why some defenders of biological diversity argue that it constitutes a strategic reservoir for the pharmaceutical industry, which must be preserved particularly in view of the fact that it is still imperfectly exploited.

Diverse animal products are also used in traditional medicine without their active components being explicitly defined. Trade in rhinoceros horns and tiger bones was especially lucrative but extremely damaging to the species concerned.

7.6.2 Biological diversity and the pharmaceutical industry

Over the last 150 years, medicinal plants have provided the pharmaceutical industry with highly effective forms of medication. New medicines have been developed by isolating the active components of medicinal plants, most of which are toxic plants. This was the case with digitalis, the source for cardiac medicine, or the poppy, which yields morphine.

Aspirin, the universal pharmaceutical product, is derived from acetylsalicylic acid, which was discovered in the meadowsweet; penicillin comes from fungi belonging to the genus *Penicillium*. Other important medicines have recently been extracted from natural products: for example, anti-tumour agents from the Madagascar periwinkle (alkaloids) and from the bark of the American yew tree (taxol).

However, the search for new molecules encounters a number of serious obstacles. Once a substance has been identified, it is not always possible to procure a sufficient amount of living material to ensure its exploitation. Thus, in 1987, the production of taxol required around 30 tonnes of yew tree bark. Since the American yew grows slowly, the tree was in danger of becoming extinct, eliciting protests from natural conservation movements. To resolve the conflict, chemists tried to synthesize the active components. In the case of taxol, they met with success, but in general, the success of such attempts remains uncertain.

Animals are also the source of pharmacological substances. Shark liver contains substances that increase the human organism's resistance to cancer. Bee's venom is used to treat arthritis, and the venom of numerous snakes is used in pharmacology.

The global pharmaceutical industry has a vested interest in biological diversity. A considerable proportion of its profits derive from drugs created on the basis of biologically active principles extracted directly from plants or else first identified in plants before being produced on a synthetic basis. More than half the drugs currently used contain a natural plant or animal extract as an active ingredient.

Pharmaceutical research seeks to profit from the vast knowledge of plants accumulated by natural healers. Ethnopharmacologists study traditional medicine and their pharmacopoeia in an effort to relate ancestral knowledge of traditional medicine to the scientific knowledge of today. But there are also other approaches to the little explored world of natural medical substances. Ambitious programmes for systematic research into new substances have been implemented by the pharmaceutical industry. Their strategy is to sift through the maximum number of species. In this way, cyclosporin, which has enabled great advances in organ transplantation by suppressing immune reactions in the recipient, was discovered in fungi.

In the recent decades, the extraordinary diversity of marine flora and fauna has stimulated scientists to explore this under-exploited environment in search of new molecules with unknown chemical properties. So far, several thousand substances have been registered, some belonging to new classes of molecules with no terrestrial equivalent. Almost one half of the marine molecules recorded in the world since 1969 have properties that are effective against tumours: cytarabine (an antileukaemic agent commercialized under the name of Aracytine) is derived from a sponge from the Caribbean Sea; bryostatin, derived from a bryozoan in the Gulf of California, is particularly promising in that it inhibits the development of solid tumours and melanomas. Major pharmaceutical companies are also interested in using neurotoxins (e.g. the venom of gastropods) to produce analgesics.

Of the thousands of molecules of marine origin identified so far, only a few dozen have commercial potential. The main reason is that many substances have been isolated in rare species that cannot be harvested in great quantities and are impossible to cultivate. Chemical synthesis is often difficult, given the extremely complex chemical structure of the new molecules.

Since the beginning of the 20th century, exploration of the microbiological world has added its share of active components to the therapeutic arsenal deriving from the plant world. This is the era of antibiotics: penicillin, tetracyclines, streptomycins, etc. In fact, nature contains an extraordinary diversity of molecular structures. These natural substances have become major international trade commodities. With their great biological diversity, many developing countries potentially possess a hitherto unexploited source of molecules coveted by large pharmaceutical companies and the cosmetics industry.

8 Genetic Resources and Biotechnology

Genetic resources are genetic material with actual or potential economic value. Agriculture, breeding, land management and industrial processes, the transformation and commercialization of agro-alimentary products – all these factors have played a role in leading humans to select and multiply a limited number of animal breeds, plant varieties and microbial strains. Their choice falls upon those that perform best towards meeting the objectives of ensuring the food supply and managing the local environment, while exploiting intrinsic properties that generate profits.

Genetic resources are a fundamental feature of biological diversity. Ancient and modern plant varieties, local breeds and their wild relatives together form the basis for maintaining or creating production systems and for modelling cultivated species to meet to different agricultural, industrial or medical needs. Genetic resources are part of the cultural and technical heritage of humankind. However, they have been partially eroded over the course of recent decades. Since the 'green revolution' of 1960–1970, varieties of high-yield crops requiring intensive use of pesticides and fertilizers have been strongly promoted, with a resultant neglect of certain regional varieties.

8.1 The Domestication of Nature: an Old Story

Ever since the emergence of the species *Homo sapiens*, plant and animal resources have been used on a daily basis and modified to accommodate epidemic or climatic circumstances. The mastery of tools and fire increased human efficiency in gathering, hunting and cultivating. They were able to roam farther, exploit their environment on a larger scale and improve the practice of domestication. Agriculture associated with ground clearing and deforestation was a great innovation of Neolithic

Biodiversity   Christian Lévêque and Jean-Claude Mounolou
© 2004 John Wiley & Sons, Ltd   ISBN 0 470 84956 8 (Hbk)   ISBN 0 470 84957 6 (pbk)

peoples. The practice of breeding cattle, goats, sheep and dogs dates from this epoch, as does the conscious selection of trees to be tended and plants to be cultivated and improved.

In western Europe, the Roman conquests and the spread of Christianity introduced hitherto unknown plants and techniques: tilling the soil, irrigation, cereal and vine cultures, swine raising, reserves for small game (rabbits), etc. This transformation was the delayed heritage of changes that had taken place in eastern Africa, Chinese Asia, India, Persia, the Middle East, Egypt and Greece at much earlier dates. In short, for thousands of years, evolving societies around the planet have derived benefit from the conscious modification of their biological environment. History also tells of attempts that failed: overexploitation and/or climatic change have sometimes destroyed freely accessible public goods and available resources. The disappearance of societies in the Sahara and the lost cities of the Euphrates valley are cases in point.

In Europe of the Middle Ages, the need to support the expansion of feudal and religious structures provided a fresh impulse to exploit the living world. To combat famines, the monks promoted ground clearing and deforestation and initiated new forms of land cultivation and breeding. The Crusaders returned from the Orient, bringing many unknown plants (vegetables and fruit trees) and animals (cats) that are now considered endemic to Europe. Imports and transformations, some replacing endemic European species, were later promoted by trade with the eastern Mediterranean and Asia. The merchants of Genoa and Venice introduced new products (spices, dyes, etc.). The natural environment changed, relationships between humans and the living world grew more complex, while demographic and economic growth continued.

At the end of the 16th century, Olivier de Serres wrote an agricultural treatise advocating prudent farming practices. His primary concern was balancing imports and exports. He advocated distributing the profits from cultivation between immediate gain and investment towards maintaining the potential for production, soil fertility and the diversity of resources. The author warned against forms of exploitation that extract benefits from the living world like salt from a mine, without bothering to give agricultural systems the means and time to recuperate. For herein resides the remarkable quality of life: that it is capable of reproducing and can tolerate withdrawals, on condition that these do not imperil its capacity for reconstruction.

The transoceanic voyages, major expeditions and colonisation of new worlds from the 16th to 19th centuries placed western Europe at the hub of these developments. New species were introduced, some of which have become indispensable to our lives (corn, potatoes, tobacco, tomatoes, etc.). Transfers to other continents also occurred (it was through humans that the rabbit acquired a foothold in Australia). Horses, extinct in America, were reintroduced by colonial settlers, together with sheep, cattle, sparrows, and a series of plants, but also smallpox and syphilis. The biological diversity of the American Great Plains was greatly transformed by these imports.

The last two centuries have been marked by a growing, gripping and global hold of humankind over biological diversity. Population growth, scientific and technological progress, the evolution of political ideas and nations, the recent dominance of capitalist economies – all derive a proportion of their success from the accelerated exploitation of the living world at all its levels of organization.

This grip can be illustrated in several ways. After the Battle of Trafalgar in 1805, English dominion of the seas effectively cut off the countries of the Napoleonic Empire from their overseas colonies. To compensate for the shortage of colonial sugar, the French government ordered extended domestication and cultivation of beetroot, altering the biodiversity of French agricultural systems and causing new technologies to be developed. At the same time, the power struggle in the Old World caused a severe social crisis in the Antilles, where biodiversity had already been ruined a hundred years earlier by the monoculture of sugar cane.

In regions all around the globe, colonial conquests, cash crops and international trade have transformed biological diversity, provoking regional ecological crises: deforestation; industrial crops; aggressive, large-scale practices in the stead of gentler ones; and the introduction of invasive species and new techniques that modify the environment.

This rapid evolution has the trappings of a frantic treasure hunt, keen on immediate profits, with little thought for tomorrow. The successes are matched by failures with possibly long-term effects. The extinction of species, the disappearance of ecosystems and aboriginal social structures are visible and foreseeable manifestations. After extending around the world, modern societies are beginning to realise that biological diversity is not inexhaustible.

Sugar . . . and slavery

Native to Polynesia, wild sugar cane was imported to India, where it is known to have been cultivated since 1200 B.C. The Assyrians, later followed by the Darius' Persians, adapted cane to their regional climates and elaborated methods for extracting and refining its sugar. Sugar was the staple good of the spice trade, which enriched the ports of the Mediterranean region. The first industrial sugar cane refinery was established in Crete (its Arabian name *Quandi* is the etymological root of sugar candy) to produce sugar for trade with the West, where it sold for a staggering price in the Middle Ages. During the Crusades, noblemen of the west conquered the island of Cyprus, where they planted sugar cane. Later, in the 15th century, the Portuguese acclimatized cane to the island of Madeira. The discovery of the New World opened up new prospects for sugar cane cultivation, most notably in the West Indies (Caribbean islands). Growing sugar cane requires intensive manual labour; the first to be enlisted were convicts and confirmed criminals. But faced with the 'negative attitude' of the Caribbean Indians (who paid with their lives for refusing to work for European farmers), producers turned to black slaves from Africa to keep the sugar plantations operating. Established by the English and French, a system of 'triangular commerce' developed among Africa, America and Europe: slaves were purchased on the coasts of Africa, shipped to the Caribbean, and exchanged for rum and sugar. During the 17th and 18th centuries, such practices constituted the backbone of global commerce. It was not until the mid-19th century that slavery was definitively abolished by the European countries. This coincided with the emergence of sugar processing on the basis of sugar beet – an unexpected result of the colonial trade barrier erected by England against Napoleonic France.

8.2   Genetic Diversity and Domestication: Creating and Selecting 'Useful' Species

Humankind has been influencing genetic diversity for a long time. Domestication is a process of coevolution of a group of animals or plants. Its dynamics are driven by the interaction of the species themselves and the choices that farmers and breeders make and repeat over time: breed selection and maintenance to obtain a particular product and/or to adapt it to the needs of a particular social system. Strategies for genetic diversity management depend upon the control exerted over the animals or plants involved. They differ from the strategies applied to wild species.

Acclimatization, cultivation, preservation and breeding are possible thanks to the consistent reproduction of living cells based on the semiconservative replication of DNA and the even distribution of hereditary matter among the daughter cells at the time of cellular division. Techniques such as cloning, micropropagation, cutting, grafting, etc., make use of these properties, enabling selected genotypes to be multiplied, in theory indefinitely (or in any case, enough to meet the demands of agriculture and industry).

A different approach is the genetic improvement of plants, animals and micro-organisms, which relies upon processes of mutation to enrich the array of available genes. Useful spontaneous mutations may be found in cultivated species and close relatives in the wild; experimental mutagenesis represents another possible means for enriching the set of available genes. After more than a century of research in crossing individuals with well-defined genotypes and analysing their descendants, geneticists have succeeded in rationalizing the empirically derived methods of traditional plant breeders and cross-breeders. It is now also possible to compose recombined genotypes, creating hitherto unknown sets of genes that are of greater interest than their parents, and to propagate them for the benefit of society.

Breeders do nothing other than make use of natural biological mechanisms for sexual reproduction (animals, fungi, plants) and genetic exchanges (bacteria) to mix and recombine DNA molecules. Mendelian and quantitative genetics were founded upon these mechanisms. Entire arrays of extremely varied genotypes (i.e. individuals) have been created by these means, both in the open field and in the laboratory. Together, they constitute a man-made enrichment of biological diversity. These domesticated species live and multiply under the control of humans, who ensure their maintenance and protect them against genetic drift and the selective pressure that they would undergo if they were in competition with 'wild' ones in a natural environment. To avoid the drift and spontaneous mutations that would normally occur in the course of maintaining and increasing cultivated population sizes, given the expected rate of diversification, collections of these specimens are conserved in a dormant state: dry seeds, freeze-dried pollen and spores, preserved sperm, somatic cells and mycelia. Administrators can extract aliquots from these reserves to check that the types conform to the catalogued genotypes, strains, local breeds, cultivated varieties or ecotypes created by genetic experimentation and selection or gathered from natural sources.

Historically and over the long term, the genetic diversity of domestic species appears to evolve through periods of depletion and periods of enrichment. The former are phases of impoverishment, when priority is given to few domestic strains, while, whether deliberately or not, the wild relatives of domestic species are abandoned and disappear. This was the fate of the wild relatives of cattle and horses. In the second phase, by contrast, farmers and breeders have focused their efforts on enriching the genetic diversity available within domestic populations. Already in the 19th century, Darwin observed that domestic populations were more varied in outer appearance than wild populations. In the case of rabbits, for example, the coats of domestic animals are much more varied than those of their wild relatives. The genetic diversity currently available in conservatories and collections is quite substantial.

The process of domestication is still continuing, and we are currently entering a new phase of genetic diversity depletion, moving away from numerous locally selected populations, and towards standardised breeds, varieties, lines, and even clones.

The primary tools for describing diversity are still the tools of demographic analysis, adapted to the methods of breeding or cultivation; genetic effective population size (N_e) is a major parameter. For populations maintained primarily under human control, genealogies provide comprehensive information on relationships, consanguinity, founder effects, population sizes and bottlenecks. Evaluated in terms of neutral, unselected polymorphism, the study of their genetic diversity provides insight into the biological histories of different populations of a species; evaluated in terms of quantitative and selected characteristics, genetic diversity sheds light on economic and social history, as well as the ways in which humankind has used genetic diversity. The tendency towards depletion that is evident today, at the beginning of the 21st century, is causing an impoverishment of breeds and varieties and a risk for the future. The situation is not identical for all species, but the FAO has rightly recommended that domestic genetic diversity be identified, inventoried, preserved, administered, evaluated and valued.

The general aim of a policy for the dynamic conservation of genetic diversity could be long-term maximization. One guide for decision making is to consider how the disappearance of a breed or variety would impact global diversity and evaluate the economic and social consequences, both for the immediate future and over the long term. In the final analysis, the choice is often between preserving endangered breeds or varieties as against improving the variability of breeds and dominant varieties. There is no hard and fast rule for making such choices . . . , but whatever route is chosen – the former, the latter, or in between the two – agreement must be reached on three management principles: decreasing variances in the population size of reproducers over successive generations; optimizing the number of reproducers (particularly for male animals); regulating the genetic flow between distinct groups.

Recent developments in the field of genetics have added two further contributions to the debate on biodiversity and the conservation of genetic resources. The first is finally to clarify the enduring controversy over the relationship between the probability of mutation and the pressure of selection. The second draws attention to time scales and rates of change in biodiversity. If the impact of human activities has the effect of accelerating the rhythm of change, then living organisms will find themselves in a perpetual race to exploit their mutator genes. Of course, they will have transformations imposed upon them, but they will increase their genetic baggage and their horizontal exchanges. This process offers rich possibilities for biotechnology to develop new products and therapies and reap new profits. It also creates biological risks, which may develop into crises and situations that are difficult to manage and sometimes irreversible, as in the case of biological invasions.

8.3   The Management and Diversity of Genetic Resources

Collections of diversified genotypes constitute valuable resources for discovering new genetic combinations to meet the needs of agricultural production. In a society that prioritizes economic profit, agro-industries competing to produce, improve and transform agricultural products enjoy a comparative advantage if they have access to these collections. Consequently, such collections of genetic resources (also called germplasm collections on gene banks) tend to accumulate mainly species that are of economic value. For research institutions, such collections also provide a unique tool for studying processes of genetic diversification and the evolution of biodiversity in situations of extreme selective pressure, such as those created by desertification, pollution or climatic changes.

Genetic resources are collected to ensure the conservation, availability and, if possible, diversification of biological material that is of immediate or future use for human purposes. Two forms of management coexist: *ex situ* management and *in situ* management.

Ex situ management means that species are maintained outside their habitats. Specimens collected during field prospecting or created through artificial selection are preserved in stable conditions, sheltered to some degree from the natural evolutionary forces of natural selection and genetic drift. In *ex-situ* management, the process of genetic enrichment is entirely controlled by humans, through a variety of techniques.

- *Seed, pollen and spore banks*: most plant species produce seeds, which are fairly easy to store. For certain species, e.g. most cereals, the seeds can be dried and preserved at low temperature (around -20 °C) without losing their viability. Some seeds can survive for a hundred years in this way.

- Another method involves *in vitro tissue cultivation*: minuscule plant parts are conserved in test tubes, where propagules are grown in a nutritive solution. This method is suitable for cloning species and storing them under decelerated growth conditions. Notwithstanding its limitations, this is the only *ex-situ* means of conserving plants that do not produce seeds or that propagate by rhizomes or bulbs. It is sometimes associated with cryopreservation, which consists in maintaining tissue cultures at a very low temperature, e.g. in liquid nitrogen (-196 °C).

In situ conservation is concerned with maintaining species' populations in natural habitats where they occur, whether as uncultivated plant communities or in farmers' fields within the context of existing agro-ecosystems. In field conditions (ecosystems, agrosystems), genetic diversity may continue to evolve under the complex constraints of natural environments. Several techniques are used.

- *Field gene banks*: plant species that do not produce seeds easily, or whose seeds cannot be stored in a 'dormant' state, can be maintained as field-grown material. Numerous cultivated species that are important for tropical countries multiply by vegetative reproduction (sweet potatoes, maniocs, yams). At the same time, sample plants are preserved in botanic gardens, arboretums and research sites. The genetic material of various species, such as the hevea tree, coconut palm, manioc, banana tree and coffee plant, is also preserved in this form.

- *On-farm conservation*: the goal is to preserve the many local varieties of cultivated plants or domestic animals that have been selected by farmers over the long term to suit local conditions or specific uses. In many countries, farmers practice conservation of genetic diversity on the farm by maintaining traditional breeds.

- *In situ conservation of genetic resources for wild relatives of cultivated plants* requires a specific approach, since most protected areas are established to preserve a famous landscape or to save a rare mammal or bird, but rarely to preserve a wild plant. Moreover, many of these wild varieties are only present in quite limited zones. Consequently, existing nature reserves are not always suitable for creating genetic reserves of wild relatives of crop plants, and specialized nature reserves are often necessary (see Figure 8.1). Priority should be given to species that do not lend themselves to *ex situ* preservation, such as the hevea tree in Amazonia, the cocoa bean and groundnut in Latin America, the coffee tree in Africa, citrus trees in Asia, etc.

The two management methods are complementary ways of optimising the enrichment of resources. There are approximately 200 public and private genetic resource banks (germplasm collections) around the world. In France, for example, they are co-ordinated by the Genetic Resources Board, which establishes management principles and reference methods. The collections are organized in networks according to plant types (cereals, conifers, etc.). On the global level, they are administered by the International Centres for Agronomic Research, executive branches of the FAO. These institutions are especially concerned with conserving and enriching tropical resources.

In terms of biodiversity, all these collections suffer from a fundamental defect: they do not represent a balanced sample of global plant diversity, since 60 per cent of the gathered specimens derive from less than 1% of living species. Full exploration of potential reserves is far from assured: botanists estimate that we eat only 3000 plant species, whereas 20 000 are actually edible!

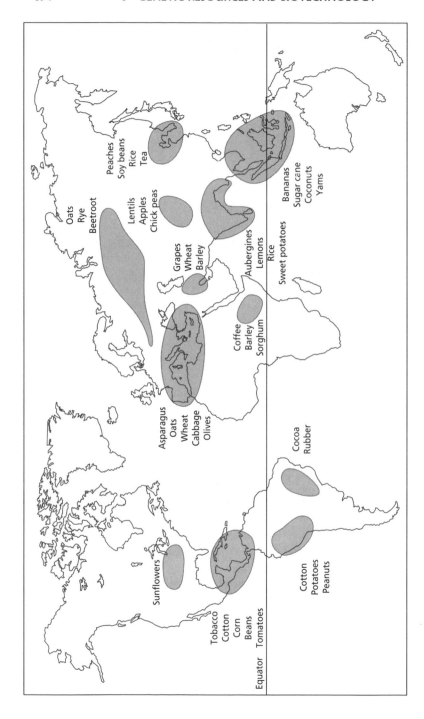

Figure 8.1   The world centres for the origin of cultivated plants are also important centres of genetic resource diversity, because they still harbour many wild relatives of major crops

Be that as it may, with a total of almost 5 million specimens, the collections of genetic resources represent an incomparable treasure. Thanks to the Green revolution, they have enabled agricultural production to keep pace with the global demographic growth of human populations (cereals – rice, corn, flour, millet – are the most striking examples).

Animal genetic resource management is a different matter: only around 40 species are concerned, and most of these are not preserved in formally structured genetic resource banks. Only salmon and bees are treated along the same lines as plants. On the other hand, sperm banks undertake a considerable amount of cryopreservation in connection with artificial insemination programmes. Genetic resource management is dynamic and integrated in the process and improvement of domestic animal breeding (cattle, horses, etc.). Conservation is merely a secondary objective behind performance and production goals. Collections, sperm banks and controlled breeding are nevertheless indispensable to maintaining the diversity of domestic races. The preservation of local breeds (sheep, goats, fowl, etc.) that are not in high demand requires adapted breeding plans and specialized herds and flocks. The financial cost is shared among public institutions, professional associations and the large number of hobby breeders.

Collections of fungi and micro-organisms serve a two-fold purpose: to provide both a systematic reference and a source of material for users. The World Federation for Culture Collections has recorded around 800 000 strains preserved in close to 500 collections, which are responsible for assigning nomenclatures. Fungi and yeast represent 44 per cent, bacteria 43 per cent, viruses 2 per cent, while the remaining 11 per cent are made up of algae, protozoa, protists and plasmids. Like other collections, they give only a biased picture of biological diversity. Over two-thirds of the micro-organisms observed by microbiologists cannot be cultivated in artificial laboratory conditions and are therefore not included in living collections (although they are found in banks storing DNA and DNA sequences).

Genetic resource collections may be a long way from fulfilling their theoretical objectives: they are expensive and difficult to maintain, and their very existence is threatened in situations of economic crisis. Yet despite all these drawbacks, they have proved their scientific, economic and social value. There is a struggle among countries, communities, industrial companies (both national and international) and individuals to acquire the rights of access to the genetic resources of today and

---

Domestic animals

Natural and artificial selection have produced thousands of genetically different breeds of domestic animals adapted to a great variety of natural environments. In agriculture, intensified breeding over recent decades has homogenised production and led to a decrease in the number of animal breeds used. According to a report by the FAO, at least 1000 breeds of animals once used were lost in the course of the 20th century, and around 2200 breeds could disappear in coming decades, i.e. one-third of the approximately 6400 breeds of mammals and birds once bred by humans. Close to half of the 2576 domestic varieties currently registered in Europe are threatened by extinction due to lack of economic profitability. On the other hand, conservation of this diversity might enable selection of animals capable of resisting different diseases, adapting to climate changes or meeting consumer demand, depending on the circumstances.

---

tomorrow. The stakes extend beyond the realm of biology. They are so large that the implementation of the Convention on Biological Diversity is turning primarily upon negotiations over statutes, rights and royalties. There is no doubt that in the future, global regulations for commerce and intellectual property will also address the domain of genetic resources.

8.4   The Biotechnological Revolution and Genetically Modified Organisms

The term biotechnology is used to designate any kind of technological application involving biological systems, living organisms or their derivatives, in order to create or modify products or processes for specific purposes.

As a process for modifying living things, biotechnology is an ancient technique. Humans have long created animal breeds and plant varieties by using methods of cross-breeding, hybridization and selection. Bio-technologies have also been used for a long time in fermentation and in the food industry, whether for aromas, colourings or other additives.

However, modern progress in the field of microbiology provides new tools that can be used to modify the living world.

8.4.1 Transgenesis

Transgenesis consists in causing part of the genetic heritage of one organism (known as the donor organism) to express itself in an organism of another species (host organism). Many potential applications of this technique result from the possibility, for example, of introducing new characteristics into an organism that it would otherwise not have been able to acquire. Thus, in principle, a plant can integrate a gene derived from a fish, bacteria or human being. But transgenesis does not generate 'unnatural' or unviable hybrids. Exchanges are limited to certain genetic elements leading to the modification of a few specific properties in the host. There is nothing very original about the operation as such, insofar as nature itself proceeds by exchanges and spontaneous genetic associations: natural plant polyploids were produced by process of hybridization; this is

---

Genomics and integrative biology

The analysis of genomes, or 'genomics' inventories the genes of a particular organism in order to study their functions. This discipline was born in the late 1980s. Structural genomics describes the organisation of the genome, maps out its sequence and compiles an inventory of its genes. The genomes of two model plant organisms, rice (50 000 genes) and *Arabidopsis* (approx. 25 000 genes) have been mapped out. The genetic information of humans (approx. 32 000 genes) has just been decrypted, following that of the yeast (*Saccharomyces cerevisiae*, 5800 genes) in 1996, a nematode worm (*Caenorhabditis elegans*, 19 000 genes) in 1998 and the fruit fly (*Drosophila melanogaster*, 13 600 genes) in 2000. Functional genomics studies the functions of genes, their regulatory methods and interactions. Thus, the field of genomics includes all aspects of the genome, i.e. the entire biological system. In the wake of these advances, a new era is approaching, where comprehensive investigation of the genome will play an integrative role in biology. These are the beginnings of integrative biology: encompassing biological phenomena from the molecule to the organism and its existence within its environment.

how virulent properties are transmitted between bacterial species, and how viruses are passed from one animal to another or to humans.

In much the same way that civil engineering relates to the techniques applied by engineers to build roads and bridges, genetic engineering comprises the set of tools and methods used to confer new properties upon living cells by modifying their genetic material. Such modifications are achieved through combination of different DNA molecules. This is why genetic engineering skills are sometimes referred to as 'DNA recombinant technology.' Genetic engineering draws upon technological advances in many scientific domains: cellular biology, biochemistry, genetics, etc. It constitutes a separate and distinct field within the domain of biotechnology.

The basic concept underlying genetic engineering is the fundamental unity of living things. Transgenesis is made possible by the universality of the vehicle for genetic information, the linear organisation of genes in the genome, and the genetic code. There are certain genes and DNA sequences that are practically identical for bacteria, plants and animals, including humans. Others are specific to particular taxa or different species.

Once they are integrated in the genome of an organism, transgenes are passed on to descendents in the same way as all other genes. For species that are of economic interest, such as major crop species, this makes it possible to introduce genes increasing resistance to diseases, pests or herbicides, or encouraging the production of new proteins. These new qualities translate into agricultural or industrial advantages, while at the same time harbouring potential risks for natural biological diversity.

8.4.2   Applications in agriculture

Selection was the traditional form of genetic manipulation in agriculture. Through selection, it was possible to create many breeds and varieties, as well as hybrids between different species. Today, these simple but time-consuming methods have been supplanted by methods deriving from molecular biology. A single gene can be extracted from an animal or plant cell and inserted into another individual of the same species or a different species, causing the latter to acquire the desired information. These living products of modern biotechnology are called GMO (genetically modified organisms) or LMO (living modified organisms).

Bio-informatics

DNA is described by the letters A, C, T and G, which represent the four types of nucleotides making up the genetic code. Genetic information on living organisms is stored in huge databases. With the accumulation of data from genome sequencing, their volume of available information is increasing at an exponential rate. What to do with this enormous mass of data? Bio-informatics is a new discipline emanating from biology and information science. Its objective is to convert the accumulation of crude data into information that biologists can use, e.g. to compare the similarities and homologies between DNA sequences from different organisms. One application is in isolating active principles for the composition of new medicines. Another is to devise and refine models making it possible to predict the functions of proteins associated with genes. Looking ahead, bio-informatics specialists envisage a model for the entire series of biochemical reactions leading to the creation of life. But we are still at the exploratory stage. . . .

Genetic transfers are especially feasible for major crops. The aim is to endow certain varieties of plants with advantageous properties: resistance to herbicides or parasites, production of useful molecules (vitamins, proteins), controlled maturation, etc. In North and South America and in China, this concerns the mass production of corn, soy, rape and cotton.

Since antibiotic-resistant tobacco was first produced in 1983, an increasing number of species has been subjected to genetic transfers. In Europe, the first transgenic experiments were carried out in 1987. The first genetically modified fruit, a slowly ripening tomato, was commercialized in 1995 in the United States, followed by virus-resistant marrows and melons. Research is also underway to improve the resistance of plants to droughts and saline soil. It is possible to produce transgenic plants that secrete a substance with a toxic effect upon insect pests. This gives the plants permanent protection and avoids the need for insecticides. But there is a risk that these same properties may be transmitted to related species and cause problems for society.

In Europe, the number of authorized GMO is extremely limited. In France, authorized cultivation is restricted to herbicide-resistant tobacco and corn, as well as corn varieties producing an insecticide protein derived from *Bacillus thuringiensis*, which renders them resistant to the

European corn borer. Such permits underlie control measures and institutions (biovigilance, see box) regulating the use of the seed.

8.4.3   Anticipating the risks of GMO

The commercialisation of GMO has engendered fierce debates between the biotechnology industry, eager to reap the potential benefits of their genetic innovations, farmers whose future is at stake, and public opinion, heated by recent blows to food safety. Unfortunately, the problem of regulating potential risks posed by GMO is shrouded in uncertainty. Nobody knows *a priori* whether these GMO will have an impact upon human health or wild species; scientists are divided as to the foreseeable consequences. The extinction, persistence and propagation of GMO are regulated, as for all other living things, by the three processes of mutation, selection and genetic drift. This is the key to the problem: some of these organisms are likely to assume a position in the biosphere that is not according to plan. The future will eventually teach scientists what they need to know about the fundamental processes of biodiversity evolution. But without scientific certitude as to the shape of the future, society, driven by its economic demands, must weigh the relationship between GMO and biodiversity in terms of risks and precaution. The harmlessness of GMO has yet to be demonstrated, and contrary opinions are being voiced. The institutions have not responded adequately to the public's questions concerning the potential risks of GMO. Before technological innovation is rejected, it must be shown to be dangerous; further research is needed.

In the case of flowering plants, the dissemination of a transgene to another species may occur through sexual reproduction – through hybridization between the cultivated plants and wild relatives with which they happen to come into contact. If the latter were also to acquire resistance to herbicides, then this would eliminate any advantage gained by producing genetically modified, herbicide-resistant species. This hypothesis has become reality in the case of rape, which can hybridize with related wild species (wild rape, wild radish), and likewise for beetroot. By contrast, there seems to be no risk for corn and soy, which cannot hybridize with any other species in western Europe.

Currently, strategies to regulate the potential risks of GMO are pursuing two routes: labelling policy and the Biosecurity Protocol.

The main objective of the Biosecurity Protocol, adopted in January, 2000, in Montreal, is to protect biological diversity from the potential risks posed by LMO. It sets conditions for cross-border movements of biological entities that are capable of transferring or replicating genetic material, such as seeds, transgenic plants and animals with new genetic combinations obtained by biotechnological means.

The Protocol also draws the distinction between LMO designed for introduction into the environment and those intended as food or for transformation into food products. In the former case, the Protocol establishes a procedure of prior agreement for each GMO import, which presupposes that the country has the necessary information to make a knowledgeable decision in the matter and is able to refuse the import in case of scientific incertitude. The latter category is subject to the national and international food and hygiene regulations (*Codex Alimentarius*). This includes oils and products derived from LMO (tomato sauce, eggs produced by hens fed on transgenic corn), which the general public mistakenly considers as GMO. Their commercialization must be registered with and approved by the trade control centre for the prevention of biotechnological risks. Thus, in the European Union, commercial products containing more than 1 per cent GMO must be labelled with details of their composition. Risks from the consumption of GMO or their derivatives involve the potential presence of an undesirable (toxic or allergenic) substance in the food and/or a possible transfer of the transgene to the microflora of the digestive system.

Biovigilance

The dissemination of GMO in the environment can disturb the ecological equilibrium. The nature and gravity of the risk depend upon the characteristics of the GMO and their environments, which require case-by-case investigation. In France, the principle of biovigilance was formally established with the Agricultural Orientation Law enacted in 1999. Its realm of application includes plants, seeds, insecticides,

continues overleaf

Biovigilance (*continued*)

fertilisers and other cultivation materials composed entirely or partially of GMO, whether dispersed in the environment or offered on the market.

Using a large-scale observation system, biovigilance seeks to discover and pursue the possible appearance of unintentional effects following from new varieties of GMO and their impact upon the ecosystem. It is especially important to observe effects upon pest populations, wild flora and fauna, aquatic environments and microbial populations, including viruses. Such research has demonstrated that there is no significant difference between the entomofauna (beetles, chrysops, Syrphidae, etc.) of non-transgenic cornfields and that of transgenic cornfields able to tolerate the European corn borer.

8.5   Property Rights and the Commercialization of Living Things

While scientists investigate the role of biological diversity in ecosystem functioning, industrial interest in genes and molecules focuses on new developments in genetic engineering and the conquest of new markets. Over a period of 50 years, the perceived status of living matter has shifted away from the concept of natural objects whose components could be investigated, but not appropriated, towards the view that novelties resulting from human activity are eligible for protection like any other original human invention.

The Convention on Biological Diversity sets out the stakes explicitly. Its first article defines its objectives as 'the conservation of biological diversity, the sustainable use of its components and the fair and equitable sharing of the benefits arising out of the utilization of genetic resources, including by appropriate access to genetic resources and by appropriate transfer of relevant technologies, taking into account all rights over those resources and to technologies, and by appropriate funding'. Biological diversity is perceived as 'green gold', particularly for developing countries considering ways to charge industrial corporations of the North for access to their resources. In this sense, the Convention on Biological Diversity can also be interpreted as a legal framework establishing the modalities for exploiting biological resources.

With the spectacular advances of biotechnology, genetic patrimony has in fact come to resemble a commercial commodity. The nations ratifying the Convention were far more concerned with the distribution of royalties from the exploitation of their genetic resources than with the conservation of species and ecosystems. We are still far removed from the notion of a common patrimony.

Thanks to genetic engineering, genes have become prime material for industry and currency for speculation. This raises the question of how biological resources may be appropriated. There are two opposing stances on this matter: one advocates free access to resources for the benefit of all (echoing the aforementioned idea of a common patrimony); the other, supported by industry, opts for a patent system designed to protect the products of genetic engineering.

A core issue of these negotiations, namely the question of intellectual property rights to living organisms, is of particular relevance to the domain of agricultural food production, where the question has been on the table for a long time. Meanwhile, conflicts over rights of access to resources and remuneration modalities have polarized the discussion, pitting the developing countries against the industrialized countries. This is essentially the reason why the agreement was not ratified by the United States.

8.5.1   The international involvement of the FAO

In the realm of phytogenetic resources, the FAO's international involvement backs the assumption that genetic resources are the common patrimony of humankind and should therefore be accessible to everyone without restriction. This notion of a common patrimony goes contrary to the ideas of private property and insistence upon national sovereignty, as they appear in the Convention. The International Centres for Agricultural Research (ICAR), which have very large gene banks, operate under the jurisdiction of the FAO and adhere to the spirit of free access, subject to international control.

The FAO also upholds the principle of rural community rights, which it defines as a right to economic compensation and/or transfers of technology in return for past, present or future contributions towards conserving and exploiting phytogenetic resources. The FAO has advocated implementation of a financial mechanism to ensure that benefits deriving from the exploitation of phytogenetic resources are fairly distributed between breeders of plant varieties and the peoples from whom the resources were acquired. However, this mechanism has remained at the theoretical stage, since it has proved impossible to control the modalities of contribution and the mechanisms of redistribution.

8.5.2   The Plant Variety Protection (PVP) Certificate

Over the last century, public and private operators have applied the techniques of selection and crossbreeding to wild or domestic species to develop 'modern' varieties, and it has been considered legitimate to protect distribution of the results by appropriate measures. In much of the world, plant varieties as such cannot be patented. Rather, they are protected by a special plant variety protection system (International Union for the Protection of New Varieties of Plants–UPOV). Enacted in 1968, this convention confers an intellectual property right called PVP. The PVP certificate enables inventors of new plant varieties to reap fair remuneration for their efforts by granting them exclusive rights to the commercialisation of that variety for a period of 15 to 18 years, depending upon the species. Nevertheless, the PVP certificate is not a patent, because what is protected is the fact of having obtained a hitherto unknown variety with new characteristics as compared to existing varieties. This system of intellectual protection allows free use of the protected product in new selection schemes. The protection actually covers the specific combination of genes constituting the variety in question – the talent and the work that the breeder has invested into assembling the genes, but not the genes themselves.

8.5.3   The Convention on Biological Diversity

The Convention upholds the principle of the sovereignty of nations over their genetic resources. The concept of universal patrimony is rejected in favour of national patrimony, where nations are free either to give or to sell their genetic resources. Access to these resources requires agreement on the equitable sharing of benefits and the technological know-how necessary for their exploitation, all of which presupposes transfer mechanisms between the parties concerned. One example for the application of these principles was the bilateral agreement established in 1991 between the government of Costa Rica and the American company

Merck. For 2 million dollars, the latter acquired the right to prospect for natural resources and collect living organisms for the duration of 2 years. In return, Merck provided access to biotechnology and their benefits. It should be noted that the CBD is not applicable retroactively. This causes problems for the International Research Centres administrated by the FAO, since access to the tens of thousands of improved varieties and wild species that they have accumulated is still free.

The protection of traditional knowledge is specified in the Convention. This involves preserving and maintaining the knowledge, innovations and cultural practices of indigenous and local communities. Thinking in terms of rights to traditional resources and community-owned intellectual rights creates new legal options for protecting local resources from privatization by the biotechnology industry.

8.6 Patents on Life: an Open Debate

The rapid progress in genetic engineering has encouraged the biotechnology sector to take out patents on living organisms, commonly known as 'patents on life'. Patents are a way to protect new varieties or products, whose creation often involves a sizable investment, and have them acknowledged as the intellectual property of their discoverer. Patents give inventors the exclusive right to exploit their invention commercially for a period of 20 years. In exchange, the inventor agrees to divulge the details of the invention to the public, without fear of being copied.

The term *biopiracy* is used to refer either to unauthorized use of the traditional knowledge or biological resources of a developing country, or to patents taken out to protect 'pseudoinventions' deriving from such knowledge, without any form of compensation to its source.

Recent refinement of patent laws in developed countries has made it possible to patent genes and living organisms. This is a real revolution, since until the 1970s, it was generally considered that organisms, like natural products, could not be patented. With the Plant Patent Act of the 1930s, the USA was the first to issue patents for plants reproduced by vegetative methods. In 1980, the Supreme Court of the United States confirmed that the distinction between animate and inanimate does not apply in patent law, but that, rather, a distinction may be drawn between natural and human inventions. The Court declared it permissible to patent transgenic bacteria that 'consume' hydrocarbons. This decision is the basis for the explicit understanding that living organisms can be patented. In 1985, the United States acknowledged the patentability of a variety of corn, followed by an oyster in 1987, and in 1988, a mouse possessing a human gene predisposing it to contract cancer. In 1988, the European Patent Office also accepted that plants could be patented; then, in 1992, the patentability of a transgenic mouse that had acquired various oncogenes. In 1998, after much debate and the rejection of a first draft in 1995, the European Union also adopted a directive on the patentability of 'biotechnological inventions'.

This means, in effect, that all living things with the exception of humans can be patented, if they are the product of interventions that satisfy the conditions for patentability: novelty, inventive activity, industrial applicability.

9 'Useful' Nature: the Value and Use of Biological Diversity

The relevance of biological diversity as an economic resource was under-appreciated for a long time. Today, we see new uses emerging, with the many applications in the agricultural and food industries, pharmacology and recreational activities, not to mention the traditional activities of gathering, hunting and fishing. Considering that we are talking about one of the greatest treasures of the planet, it is paradoxical that we tend to expect free access to it. Two major reasons have brought about a change of attitude.

- Article 1 of the Convention on Biological Diversity makes explicit reference to 'the fair and equitable sharing of the benefits arising out of the utilization of genetic resources.'

In this context, it is appropriate to consider what genetic resources really can be shared. Addressing the problem of economic evaluation would appear to be a necessary preliminary to any discussion of the distribution of riches.

- According to the economists, whatever is priceless is also without value. Some have advanced the hypothesis that protection of biological diversity would be a credible undertaking only if it were possible to demonstrate the economic benefits of specific public or private decisions in conservation matters.

This raises two questions: what values can assigned to biodiversity? and by what methods can they be measured?

Biodiversity   Christian Lévêque and Jean-Claude Mounolou
© 2004 John Wiley & Sons, Ltd   ISBN 0 470 84956 8 (Hbk)   ISBN 0 470 84957 6 (pbk)

9.1   Benefits and Services Provided by Ecosystems

For a long time, nature was perceived as an inexhaustible source of free resources – some biological, others non-biological. But over recent decades, people have noted the growing scarcity of these resources and become more conscious of the economic importance of the goods and services that nature provides to humans. *Goods* are products that we buy or sell. Their monetary value is a function of the market. Timber, fish or mushrooms belong to this category. But society also depends upon services rendered by ecosystems, such as water filtration and climate regulation. These services are more difficult to evaluate in monetary terms. All these benefits and services have been inventoried (Table 9.1).

Table 9.1   Typology of benefits and services as well as functions fulfilled by ecosystems (according to Constanza *et al.*, 1997)

Benefits and services	Ecosystem functions	Examples
Gas regulation	Regulation of atmospheric chemical composition	CO_2/O_2 balance
Climate regulation	Regulation of global temperature, precipitation, and other climatic processes	Greenhouse gas regulation
Disturbance regulation	Ecosystem responses to environmental fluctuations	Flood control, drought recovery, storm protection
Water regulation	Regulation of hydrological flows	Provisioning of water for agriculture (irrigation) or industry (such as milling)
Water supply and purification	Storage, selective filtering and retention of water	Provisioning of water by watersheds, reservoirs and aquifers; soil percolation
Erosion control	Retention of soils within an ecosystem	Prevention of erosion by wind, surface runoff, or other processes
Soil formation	Soil formation processes	Weathering of rocks and accumulation of organic material
Nutrient cycling	Storage, recycling, processing and acquisition of nutrients	Nitrogen fixation, phosphorus and other nutrient cycles

Table 9.1 (*continued*)

Benefits and services	Ecosystem functions	Examples
Waste treatment	Recovery of mobile nutrients, and removal of excess nutrients and compounds	Pollution control, waste treatment, detoxification
Pollination	Movement of floral gametes	Provisioning of pollinators for the reproduction of plant populations
Biological control	Trophic–dynamic regulations of populations	Keystone predator control of prey species
Refugia	Provision of habitats for resident or transient populations	Nurseries, habitats for migratory species, etc.
Food production	The portion of gross primary production extractable as food	Production of fish, wild game, fruits, grains, etc.
Raw materials	The portion of gross primary production extractable as raw materials	Production of logs, fuel, fodder
Genetic resources	Sources of biological materials and natural substances	Medicine, genes for resistance to plant pathogens, ornamental species, etc.
Recreation	Providing opportunities for recreational activities	Ecotourism, sport fishing, and other outdoor activities
Culture	Providing opportunities for non-commercial uses	Aesthetic, artistic, educational, spiritual or scientific values of ecosystems

9.2 Theoretical Bases for Assessing the Economic Value of Biological Diversity

In academic economics, value is determined through exchange on the market and quantified in terms of price. The price reflects both the cost of production of the product and, given the choice between several products, the preference attached to it by the buyer. The destruction of a forest by the timber business falls into the market domain insofar as certain species are in high demand and some may have a high economic value. But equally, the destruction of that same forest deprives humans of other resources (fruits, mushrooms, medicinal plants, firewood, etc.), as well as services rendered by the forest: regulating biogeochemical cycles (carbon storage, for example), producing oxygen, and providing shelters and habitats for other species. The loss of all these goods and functions, in addition to the cost of reforestation, are not reflected in the market price of timber. The market therefore undervalues the price of wood by ignoring the other goods and services provided by the forest.

Thus, market prices are not reliable indicators of social cost, because they do not capture all the effects of the use of biological resources. Many natural outputs contribute to our quality of life and support our market economy but are without formal markets and therefore without prices. However, the fact that biological resource benefits are not priced does not mean that they lack value, only that market indicators of their value do not exist. Ecological economics was created to deal with this problem and look at economics from an ecological perspective. An economic analysis must account for non-priced benefits and costs, as well as those directly observed and measured in market prices.

9.2.1 Use value and non-use value

One way to begin is to assume that biological diversity has an *instrumental* or *utilitarian value* and serves to satisfy the needs of society. Another approach is to recognize that biological diversity has an *intrinsic value*, above and beyond its utility.

In other words, there is a value that is linked to a use and/or a market (e.g. a fish that is caught is sold for a certain price, as are mushrooms that are gathered); beyond that value, environmental economists see another value, independent of the market or direct utility: intrinsic value, for example the value that a plant or animal has as an end in itself, which can be estimated on the basis of society's willingness to pay for the preservation of an endangered species. This view turns upon the recognition that life itself has a value above and beyond financial speculation.

Thus, on the one hand, we have values estimated according to the use made of biodiversity (*use values*), and on the other hand, values based on criteria other than the market and utility (*non-use values*).

Utility can be direct or indirect. *Direct use values* correspond for example, to benefits derived from the production of foodstuffs, consumption in the form of hunting, gathering, fishing, extraction of industrial and pharmaceutical molecules, or else the pure pleasure of experiencing flora and fauna (green tourism). Indirect utility values essentially derive from ecological functions: e.g. water filtration, regulation of biogeochemical cycles, etc.

While *non-use values*, as opposed to use values, are not reflected in market prices, many natural assets are nevertheless assigned a value, whether it be religious, philosophical, moral, cultural, or even economic, in spite of the fact that they have no actual price. The idea that the value of natural assets does not depend solely upon their immediate use has given rise to various concepts of non-use value that measure willingness to pay an economic agent to preserve the environment and its natural assets.

- *Option value* measures the willingness to pay for the preservation of a natural asset with a view to anticipated benefits in the future. Option value is like an insurance premium.

- *Bequest value* is defined as the willingness to pay to preserve a natural asset with a view towards its use by future generations.

- *Existence value* is the psychological value a person enjoys from just knowing that a species or an ecosystem exists. It measures the willingness to pay regardless of any anticipated use in the future. An agent may agree to pay for the preservation of pandas without actually hoping to see the day when they are set free. Existence value embodies the pure concept of non-use, while option value and bequest value reflect eventual or probable use.

It is much more difficult to formalize methods for evaluating non-use values than for use values, which are reflected directly, if only partially, in economic transactions. The most traditional approach is to calculate the transportation and equipment costs paid by the consumer in order to gain access to biological diversity (nature parks, hunting, fishing). Another approach consists in creating a fictive market on the basis of surveys. Called the *contingent valuation method*, it consists in asking individuals whether and how much they would be willing to pay to preserve an element of biological diversity, whether it be a species or an ecosystem.

9.2.2 Economic goods versus free goods

To do an economic valuation of biological diversity, it is necessary to distinguish between economic goods and free goods. The former are scarce goods that can be appropriated, leading to market exchanges. The latter are free and abundant, accessible to and usable by all. By definition, only economic goods are considered by the market. Free goods are evaluated using the externality concept: the activity of an economic agent has consequences for the well-being of other agents, even though there is no commercial or financial exchange between them. Thus, a chemical firm may pollute a river, leading to a decrease in well-being or benefits for consumers situated further downstream, without the price of the product marketed by the firm reflecting this situation. In other words, some costs linked to the activity of an agent escape the sanction of the market; these hidden costs are borne by society and not by the agent responsible for the damage. Such costs, called *external effects* or *externalities*, correspond to the difference between social cost (to the community as a whole) and private cost. Mechanisms such as taxes or subventions have been proposed to offset these extra-market effects.

9.2.3   Appropriation and/or free access to biological diversity

Under what circumstances could biological diversity management conceivably operate? For some, non-domesticated biological diversity is a collective good that actually suffers from not being appropriated and managed for sustainable use over the long-term. As they see it, the absence or poor definition of property rights is the primary cause for the observed erosion of biological diversity. Uncontrolled access to public goods may cause competition among actors eager to appropriate a maximum amount of goods in a minimum amount of time, quickly leading to overexploitation. Fear of such a situation, known by the name of 'tragedy of the commons', has led some economists to suggest that resources should be privatized or that public authorities should establish a framework to control their accessibility. The Convention on Biological Diversity also circumscribes the debate over biodiversity as a choice between two alternatives: public versus private property.

But the question whether property rights should be public or private is a controversial one. In many societies, access to common resources and control of access are often regulated by customs, myths, hereditary rights or traditional regulations. Here, common property is not necessarily synonymous with open access. However, many developing countries have adopted land and resource management models from the western hemisphere. Emphasizing the role of government in resource management, such models often supplant traditional models for regulating access to resources without actually fulfilling the stated purpose of protecting these resources. In Africa, for example, centralized governmental management of inland fisheries, based on fishery models developed in European countries, is far less effective than management by local communities on the basis of traditional practices. Some experts hold that the essential role of community know-how in biodiversity conservation must be reassessed with a view towards achieving sustainable development.

9.3   Putting a Price on Biological Diversity

Since the early 1990s, conservationists have generally gone along with the strategy of promoting and applying economic arguments as a way of catching the attention of politicians and managers. Can the prospect of gains make people act reasonably? To put the question in another way: will countries with the prospect of deriving benefits from their biological diversity make the effort to protect it and avoid waste? This, in a nutshell, is the challenge of the economic approach, which consists in encouraging developers to consider biological diversity not just as a gift of nature but as a valuable resource to be preserved. Economic mechanisms must be developed to correct for the failures of the market system and revaluate the 'true' price of goods, so as to include the cost of environmental damage resulting from their production or consumption. As applied to genetic resources, this principle leads to the possibility of patenting living things, to the recognition of intellectual property rights and the introduction of royalties (see Chapter 8).

The language of economics has become the negotiating language for harmonizing divergent interests and orientating decisions on trade-offs and conflicts. At least in theory, economists have thus been given the role of mediators among scientists, the conservation movement, decision makers and stakeholders.

The market is a tool at the disposal of society to achieve its given objectives. It is not up to the market to work out priorities for biological diversity conservation; these are rather for the governments to establish through their choices in developmental matters. What the science of economics can, in theory, do is to contribute towards determining the most effective way to apply the criteria of sustainability and help to develop equitable approaches towards achieving these conservation objectives in today's market societies.

9.3.1   What is the overall value of ecosystems?

Ecosystems and their components are at the same time natural capital and a source of services in terms of both current uses and possible future uses. Defined by supply versus demand, services may involve recreational value (walking, swimming, fishing, hunting, gathering), ecological value, or value as a factor of production (energy, water resources, biological resources, etc.). The concept of 'natural infrastructure' has been proposed to characterize the major role of the services provided by ecosystems.

A team of ecologists, economists and geographers attempted an economic evaluation for all the services annually provided to humanity by all global ecosystems taken together. They identified seventeen categories of services provided by terrestrial and aquatic ecosystems (Table 9.2). Estimations based upon a compilation of available environmental economic data yielded figures ranging from 16 000 to 54 000 billion US dollars per year, more in any case than the gross national product of all the countries in the world.

Comparing these estimates for the different types of ecosystems: the median values in dollars per hectare and year vary from 577 for marine systems to 804 for terrestrial systems, 969 for forests, 8500 for lakes and rivers, and 14 785 for wetlands. Proportionately, the oceans contribute

Table 9.2   Average global value of annual ecosystem services (according to Constanza *et al.*, 1997) Some ecosystems, such as deserts, tundras, glacial and urban zones, are not included in this evaluation

Ecosystems	Surface area (10^6 hectares)	Relative value (US$ /hectare/year)	Total value (in $10^9$$ per year)	%
Oceans	33 200	252	8381	25.2
Coastal	3102	4052	12 568	37.8
Forests	4855	969	4706	14.1
Grasslands	3898	232	906	2.7
Wetlands	330	14 785	4879	14.7
Lakes and rivers	200	8498	1700	5.1
Cropland	1400	92	128	0.4
Total value of the biosphere			33 268	100

63 per cent to global value, wetlands 14.5 per cent and forests 14 per cent. In terms of goods and services, the food cycle contributes 50 per cent of the annual total. Human food production and water supply represent only 4 per cent and 5 per cent, respectively.

This study gave cause to question the methods used for evaluation. The authors frankly admit that the margins of error are very large. The key point is to strike people's imagination with the orders of magnitude involved, which can leave none indifferent. These evaluations are more-over an attempt to demonstrate that the conservation of 'ordinary' ecosystems deserves as much attention as the conservation of areas considered of great ecological value from the economical point of view.

9.3.2 'Natural infrastructures'

An ecosystem function that may provide us with a service has a non-commercial value, since there are no standard measures for its produc-tion cost or market demand on the economic level.

The idea of comparing ecological systems to 'natural infrastructures' first appeared in conjunction with the evaluation of public policy on wetlands. This concept proceeds from a fundamental observation: wet-lands fulfil functions that are comparable to those fulfilled by artificial infrastructures, such as water purification plants, or dams erected to hold back floods. A purpose of this analogy is to establish a relationship between two spheres of thought and action that affect the management of wetlands independently of one another: management of water as a physical resource; and management of and by natural environments. For wetland conservation, the conclusion is obvious: undisturbed wetlands provide a service and fulfil functions free-of-charge, whilst artificial infrastructures fulfilling similar functions, for example purification plants or dams, entail a cost to society.

If the services provided by ecological systems were paid for in line with their contribution to the global economy, prices would clearly be very different from what they are today. But how can this idea be translated into practice? In 1996, the city of New York invested 1 to 1.5 billion dollars in natural capital, with the prospect of economising 6 to 8 billion dollars in 10 years. The city's water supply comes from the Catskill Mountains. For a long time, self-purification through soil filtration was sufficient to meet the quality standards of the Environmental Protection Agency (EPA). But with growing amounts of wastes, fertilizers and pesticides, natural purification was no longer sufficient, and the city faced a critical choice: either to construct a purification plant for 6 to 8 billion dollars plus an annual 300 million dollars in operating and maintenance costs, or to restore the integrity of the Catskill Mountain ecosystems. In this case, the investment in natural capital consisted in buying the land comprising and surrounding the watershed in order to restrict its use.

9.4   Uses for Biological Diversity

In addition to the genetic resources discussed in the previous chapter, humankind has many other uses for biological diversity, some of which can be evaluated in economic terms.

9.4.1   Use of living resources for food

The practice of extraction from the natural environment (gathering, fishing, etc.) goes back a long way in the history of humanity. Such activities are still important, albeit more diversified than they used to be. They are pursued in many different forms around the world, depending upon the regional state of economic development.

In many countries, wild and semi-wild plants contribute towards providing for the food and health of populations. These include leaves, roots, tubers, fruits and mushrooms. Some are valued as highly desirable foods, others are an indispensable source of vitamins and minerals in the diet of rural populations surviving on subsistence agriculture. Various studies have demonstrated the important role of such plants in nutrition, but their real contribution to the local economy is often nebulous, in as much as these goods do not appear in household accounts. Wild fauna also contribute greatly to feeding humanity. Many rural societies, particularly in tropical regions, consume various invertebrate species: insects (termites, grasshoppers, etc.), caterpillars, molluscs. Many vertebrates (mammals, birds, reptiles) are hunted for their meat, and fishing in ocean or inland waters is the primary source of protein in the world. Annual extractions from aquatic environments total an estimated 100 million tonnes, with an estimated overall value of 60 to 100 billion Euros. In many cases the levels of extraction are close to the limits beyond which resource regeneration is no longer possible. Actually, a number of natural resources are already overexploited.

From French wine to Coca Cola

It was a French pharmacist, Ange Mariani, who first hit upon the idea of decocting coca leaves in red wine, a kind of forerunner of Coca Cola. Proffered for sale in 1884, his 'coca wine' was a luxury good with great success in high society. In 1886, John Smith Pemberton, a pharmacist in Atlanta (USA), commercialized what he called 'French Wine Cola' as both an excellent tonic and an ideal stimulant. Soon after, Pemberton succumbed to the pressure of the temperance leagues and offered an alternative without alcohol, replacing the wine with lemon juice and adding carbonation, and the Coca Cola brand was born. A few years later, Pemberton sold his invention to another pharmacist, Asa Griggs Candler, who promoted the product. Thus began its long commercial ascent. . . .

9.4.2 Extractive products

The term extractivism refers to the commercial exploitation of non-wood forest products, such as fruits, gums, resins, oils, fibres, etc. In the Amazon region, the commercial value of 12 different forest products (fruits, rubber, etc.) is estimated at 420 US dollars per hectare and year. But such estimates are highly dependent upon the proximity of markets, transport facilities, as well as possible outlets in industrialized countries.

The history of rubber is well known, together with the economic competition unleashed by this product, so essential to Western industry. Even if it is now possible to produce synthetic rubber, latex derived from hevea has certain properties that are not found in its industrial counterpart and continues to be much in demand. Currently, around 5 million tons of natural rubber are produced annually, mainly by Asian countries, as compared to 10 million tons of synthetic rubber.

The rubber saga

One South American plant that played an important role in the industrial development of the past century is the rubber tree, also known as the 'wood that cries.' Discovered by the earliest conquistadors, it was

continues overleaf

The rubber saga (*continued*)

used by the Aztecs to make elastic balls for *tlachli*, a very popular game with strong religious overtones that usually culminated in human sacrifice.

The Spaniards learned how to impregnate barrels and tissues by greasing them with latex, but because it lost its qualities in the cold, its use in Europe remained limited. The process of *vulcanization* (heating rubber with sulphur) was discovered accidentally in 1839 in the United States by Charles Goodyear. By stabilizing the physical properties of latex and rendering it impervious to changes in temperature, this process opened the door to its industrial future. Demand grew, inciting particularly intensive exploitation at the end of the 19th century, not only in the Amazon region of Brazil, where the seringueros extracted rubber from hevea trees, but also in Africa, where it was extracted from lianas and euphorbias.

With the invention of the rubber tyre and the development of the bicycle and later the automobile, the amount of rubber harvested was no longer sufficient. In the early 20th century, the Amazon region which had been the cradle of this resource gradually lost all benefit from its exploitation, as large plantations of *Hevea brasiliensis* were created first in Asia and later in Africa. Mass production in these two continents partly destroyed their original economies. Eventually, all natural rubber producers experienced difficulties through the competition of synthetic rubber.

Cotton, used in the textile industry, also has a long and complex history. Domesticated around 5000 years ago in Persia (from whence it reached India), Peru and Mexico, it was a luxury good until the 18th century, when cheap, American cotton became popular in Europe. Other textile products such as hemp, linen, wool, silk and leather are also derived from plants and animals. These products have played an important role in the global economy and are still largely in use today, despite the competition of synthetic products.

The wicker, textile and rope-making industries also use a wide range of natural products. Approximately 600 species of Asian rattan are the basis for large industries in China, India, Indonesia, Thailand and the Philippines. Of all forest products, international trade in rattan is second only to wood. Commerce in leather and hides is also considerable, with both crafts and industry profiting from the diversity of species used.

9.4.3   Timber

Timber trade is an important economic activity on the international level. Forests cover slightly more than 3400×10^6 hectares, or approximately one-quarter of the overall dry land surface of the planet. In the developing countries, wood is mostly used for domestic purposes (firewood), as well as for construction. In developed countries, great quantities of wood are consumed by industrial applications (e.g. for paper pulp). Most commercial timber comes from temperate countries (Canada, Finland, Russia, USA), where production is well organised. But demand for precious woods is also great, and forest resources constitute a rare source of foreign revenue for many tropical countries. They are sometimes tempted to exploit this heritage in an excessive manner, given the time forests require to regenerate. Lacking adequate management, many forest environments are currently suffering extreme degradation due to overexploitation.

9.4.4   Biotechnology and industrial applications

Industry has a direct interest in certain elements of biodiversity, such as micro-organisms, molecules, genes, etc.: the food and agriculture sectors work with the diversity of cultivated plants (see Chapter 8); the pharmaceutical sector is more interested in wild species (Chapter 7). In all areas, the application of biotechnologies on an industrial scale is a strategic economic challenge.

Industrial microbiology uses the metabolic and enzyme functions of micro-organisms for two main types of transformations:

- fermentation of raw materials in agriculture and food production (wine-making, brewing, cheese-making, etc.) or in depollution. The micro-organisms used in these processes are found in foods or the environment;

- production or modification of very different molecules (enzymes, antibiotics, hormones, aromas, etc.). In this case, the culture is produced in a contained environment, and the cells are destroyed when the production process is completed.

Strains of micro-organisms are initially selected from the natural environment and may be genetically improved to produce more efficient strains that, as opposed to higher plants and animals, rarely require crossbreeding. Improvement is based on studying spontaneous mutations or mutations caused by mutagen agents. Since the end of the 1970s, techniques of molecular biotechnology have also been applied to adapt genetically modified micro-organisms for industrial production. Such strains are only used for that purpose and not to produce food.

In the health domain, genetically modified organisms have been widely used since the late 1970s to produce recombinant proteins (obtained by genetic recombination). Bacteria have been genetically modified to synthesize molecules such as human insulin for the treatment of diabetes, growth hormones, and erythropoetin, which stimulates the production of red blood corpuscles, etc.

Most of the insulin used today derives from recombinant bacteria rather than from porcine pancreas. Since 1986, bacterial production of growth hormones has enabled risk-free treatment of children suffering from certain forms of dwarfism. Former treatments, using hormones extracted from the hypophyses of cadavers, entailed the risk of contamination with the agent responsible for Creutzfeldt–Jakob disease.

The vaccination against hepatitis B uses a recombinant vaccine produced by yeasts or cultures of genetically modified cells. Plant trans-

---

Bacteria and tradition: cheese-making

The variety of fermented foods such as farm cheeses, but also beer, is a product of the diversity of the micro-organisms used. Each strain produces its own particular aromas, and the more micro-organisms there are, the more aromas we have. . . . But we are still unable to describe, much still less control all the events leading to the final product. Today, the preservation of traditional food production must be reconciled with the requirements of public health. For example, there is a need for better control of microflora during fermentation in order to reduce the risk of contamination with *Listeria*. Such control is more difficult for products involving the intervention of dozens of strains than for industrial products that use only a few select strains, with less tasty results! There may be hope for improving control over microbial diversity during the course of fermentation: scientists are working on 'DNA chips' that will make it possible to follow the evolution of microflora in real time.

Animals as drug factories?

The pharmaceutical industry is perhaps at the dawn of a new revolution. A sector of biotechnological research is focused upon the use of genetically modified animals to produce medicines and nutrients. In theory, the principle is a simple one: the genetic heritage of an animal is modified to make it produce a medically useful product in its milk, blood or urine. Species such as goats, sheep, mice, rabbits, cows or chickens are already being exploited by the industry. These organisms compete with cellular cultures to produce complex molecules such as hormones, growth stimulants, antigens and antibodies at much lower production costs.

---

genesis offers further possibilities for producing medicines: tobacco can be genetically modified to produce haemoglobin, for example.

Depollution processes constitute a new field of application for biotechnologies. Some plants and micro-organisms are capable of proliferating in polluted soils and degrading pollutants, or even absorbing heavy metals in a process known as *phytoremediation*.

- Phytodegradation consists in accelerating the degradation of pollutant organic compounds (hydrocarbons, pesticides, etc.) by means of plants. Degradation may take place outside the plant, through the activity of micro-organisms present in their root environment (rhizosphere), or else within the plant, where compounds are first absorbed and then degraded in the cells.

- To extract metals from polluted soils, there are plants called hyperaccumulators that are capable of accumulating more than 1 per cent metal in their tissues. The plants can then be incinerated in such a way as to recuperate the metals that they have accumulated.

In 2000, researchers succeeded in producing a transgenic plant capable of transforming forms of toxic mercury into less harmful forms. They introduced a bacterial gene into *Arabidopsis thaliana* and demonstrated that this genetic improvement enabled the plant to survive in soil contaminated with methyl mercury. The plant's nutrition and respiration processes gradually clean the soil and can reduce its degree of pollution to a mere 2 per cent of its original level.

9.4.5   Ornamental animals and plants

Biological diversity has long been an object of recreation and leisure. The ornamental plant industry is flourishing, and many tropical species have been introduced to Europe to satisfy the demand of collectors or public curiosity. The number of plant species cultivated for ornamental purposes is far higher than the number of plants used in agriculture. There are constant innovations in this domain, and new species derived from wild forms or hybridizations regularly appear on the market.

Ornamental plants are also frequently genetically improved to produce more attractive strains, such as with larger or more beautiful flowers. It is said that blue roses are soon to appear on the market: a gene responsible for the colour blue has been isolated in the petunia and transferred to the rose to produce the desired colour. How much longer for the mythical black rose that horticulturists have been dreaming about for centuries? Over 5000 orchid species are traded internationally and an equally large number of cactus varieties. Whilst many of these species are now cultivated, a large proportion is still extracted from natural surroundings, potentially leading to their extinction is some areas.

There is also a large volume of live animal trade, whether for pets, zoos, aquariums, research activities (primates), etc. Aquarium fish are also in high demand from North American and European collectors, who may sometimes pay high sums for rare species. Certain products such as ivory, tortoise shell, snake or crocodile skin, many mammal furs, bird feathers, etc. have symbolic, cultural or decorative uses, or are worn as clothing. Because trade in these products leads to large-scale massacres, endangering the survival of a number of sought-after species, they are subject to increasingly strict controls. The same is true of the market created by insect and shell collectors. Demand is particularly high for spectacular species, often threatened by extinction if they are too intensely exploited.

9.4.6   Ecotourism

Ecotourism has become a new industry. Benefits derived from interest in biodiversity, whether observing wild animals or enjoying the attractions of beautiful natural landscapes, is an especially important source of revenue for some countries. They have developed tourist industries based on turning their natural patrimony, as such, to profit. Kenya is a good example. In some cases, nature parks, trekking, etc. attract such a large number of visitors that there is cause for concern as to the consequences of high frequentation for the preservation of the sites. Thus, tourism is also responsible for its share of ecological problems around the world, and 'ecologically aware' urban dwellers are themselves a menace to biodiversity.

10 The Conservation of Biodiversity

'Environment' is not a term we use amongst ourselves. We just say that we want to protect the whole forest. 'Environment' is what others call it, it's a White buzz-word. What you people call 'environment' is what is left over after you have destroyed the rest.

Chaman Yanomani, quoted by Descola, 1999

The conservation of biological diversity, its sustainable use and the equitable sharing of its benefits, are the fundamental objectives of the Convention on Biological Diversity. The reasoning behind this Convention and its ratification by the large majority of nations is relatively simple. It stems from the recognition that the direct impacts (over-exploitation, destruction of habitats, etc.) and indirect effects of human activities upon natural environments constitute a threat to the future of biological diversity, the renewal of resources and, more generally, to the conditions for life on Earth. Urgent measures are therefore necessary. The declared objectives of the Convention are at the same time highly ambitious and extremely vague: to foster sustainable development, while protecting and using biological resources, without reducing the diversity of species or destroying habitats and major ecosystems. The Convention contents itself with giving general directives, leaving to each country the task of taking appropriate measures in accordance with its particular geographic and social context. This has resulted in a large variety of approaches towards devising national policies, especially fraught with difficulties in realms where international economic competition imposes tight constraints.

Biodiversity   Christian Lévêque and Jean-Claude Mounolou
© 2004 John Wiley & Sons, Ltd   ISBN 0 470 84956 8 (Hbk)   ISBN 0 470 84957 6 (pbk)

The concept of protecting nature originated in the modern Western Hemisphere. It presupposes a clear duality between two ontological domains: human beings on the one hand, and non-humans on the other. In this context, nature is considered an autonomous domain, distinct from social activities. As the masters and owners of natural resources, humans naturally feel vested with the mission of assuring its preservation. But not all people share this conception of nature. There are many, for example, who confer social attributes upon plants and animals. Such cultural differences in environmental perception are symptomatic of the misunderstandings that arise between nature conservation movements and indigenous populations.

It seems obvious that if human activities are the immediate cause for the erosion of biological diversity, then the solutions and remedies for the problem must lie in the realm of social behaviour. In other words, the conservation of biological diversity is contingent upon choices made in economic development issues at both national and international levels.

The terms conservation and protection cover a large variety of practices. They may be used interchangeably or with different meanings, depending upon the country and speaker. This adds a certain amount of confusion to the debate. We propose the following definitions:

- *conservation* is an approach that considers the long-term viability of ecosystems within the context of resource and environmental management projects. Conservation involves a concept of protection that does not prevent humans from intervening in natural processes; it is rather a philosophy for managing the environment without resultant waste or depletion.

- the term *protection* is reserved for operations aimed explicitly at safeguarding environments or species endangered by human activities. The emphasis is upon defending specific ecosystems.

10.1   Why Protect Biological Diversity?

For centuries, scientists accumulated knowledge about nature without concerning themselves with the conservation of natural systems and their biological diversity. Nature was a seemingly inexhaustible reservoir, providing humans with everything they needed, whilst at the same time offering vast spaces for the disposal of waste and pollutants. During the 20th century, this attitude underwent considerable change:

- The European societies of the late 19th century tried to encourage a more rational exploitation of nature's riches. The object was to maintain conditions favourable to the regeneration of living resources so as to ensure their continued exploitation: preservation rhymed with production.

- This productivist approach led by reaction to the first ecological awareness of nature. Essentially protectionist, its philosophy was to preserve the *status quo* of certain elements of 'wild' nature. The emphasis was on conserving pristine and inviolable natural domains, sanctuaries valued as landscapes or for their flora or fauna, 'natural monuments.' This is how natural reserves and protected areas were created in many countries. Humans were considered as threats and generally excluded.

- Since the 1980's, attention has turned to the economic value of biological diversity, both as a source of genetic resources for agriculture, and for its industrial uses (new molecules for the pharmaceutical industry, biotechnologies, etc.). In this context, biological diversity is seen as a potential source of revenue, in particular for the developing countries, providing an *in fine* justification for interest in its conservation. If we fail to take the necessary measures, we shall lose the opportunity to derive profit from the potential benefits that biological diversity may bestow upon humanity.

- Lastly, it is now recognised that biological diversity plays a significant role in maintaining the major equilibriums of the biosphere. Biological diversity is involved in the water cycle and the major geochemical cycles, including the carbon and oxygen cycles. It contributes to the regulation of the physical/chemical composition of the atmosphere, influences the major climate equilibriums, and thus impacts the conditions of life on Earth. All ecological functions are a product of the complex relationships among living species.

The conservation of biological diversity is structured around two distinct but converging traditions:

- *resource management*, which implicitly acknowledges that the protection of 'useful' species is necessary for economic development. Biological diversity has an economic value; it is considered a treasure to be exploited and turned to profit. It forms the basis for human nutrition. It provides the raw materials for the agricultural and food industries, the pharmaceutical industry, the perfume industry, etc.. From our current perspective, biological diversity offers promising prospects for profit-making in the realm of biotechnology, especially considering the potential of micro-organisms, which still represent a largely unexplored world. Another source of revenue worth mentioning is ecotourism, appealing to urban dwellers keen to experience nature and observe wild species in their natural environment.

- an *ethical perception of nature*, holding that any and all disappearance of species is a loss and demanding maximum protection of biological diversity. The Conference in Rio and its debates on the Conservation of Biological Diversity clearly showed that there is a moral dimension to this question, an extension of the philosophical debate on the relationship between humans and nature. The extinction of species confronts humans with the fundamental moral problem of their relationship with other forms of life and their responsibility for preserving its diversity. As Hans Jonas put it, the question is whether one generation or one people has the right to appropriate and eliminate a large number of species that evolved over hundreds of millions of years and what is the extent of their responsibility. Do we not have the duty to bequeath to our descendents a world equivalent to that which we inherited ourselves?

The principle of responsibility

For a long time, Western societies did not think of the environment in ethical terms. Only around the 1980's did we begin to accept that our relationship with nature entails an ethical dimension. Hans Jonas was one of the precursors of this approach; in simplified terms: modern humans have such a powerful technological hold over nature that they are in a position to endanger the future of the world. Scientific and technological progress may dangerously undermine the major equilibriums of the biosphere, compromising the quality of human life and the very survival of future generations. Technology cannot be corrected by technological means. Solutions must be sought beyond the realm of rational science, necessarily invoking an ethical principle, i.e. a general theory of political, moral or legal norms to guide human actions.

Where science fails, ethics intervene. This is what Jonas calls 'the heuristics of fear.' The impending danger tells us that the survival of humanity is at stake, and we are under obligation to protect it by taking appropriate measures to avert catastrophe. Thus, humankind becomes responsible for its own future (the principle of responsibility), vested with the mission to safeguard the survival of humanity, since remaining passive would endanger its survival. Humankind today has a responsibility towards future generations. We must bequeath them a communal patrimony with access to sufficient natural resources in order that they, too, will be able to lead a decent existence.

10.2   Approaches to Conservation

The practical implementation of conservation has aroused considerable, often passionate debate as to the most appropriate modes of action. There is only one obvious conclusion: there is no simple, universal solution. Actions are often undertaken out of urgency, and in the long run, nothing is ever entirely satisfactory.

10.2.1   *In situ* and *ex situ* conservation

One customary practice is *in situ conservation*, which consists in maintaining living organisms in their natural environment. To conserve individual species, some effective approaches are: enacting legal protection for the endangered species; improving management plans; and establishing reserves to protect particular species or unique genetic resources. This kind of conservation enables plant and animal communities to pursue their evolution, whilst adapting to changes in their environment, and comprises a large number of species, without requiring preliminary inventorying. However, *in situ* conservation is not always possible, because many habitats are already seriously disturbed, and some have even disappeared entirely. In such cases, the alternative is *ex situ* conservation, which consists in preserving species outside their natural habitats. This is one of the roles of botanical and zoological gardens; other methods such as gene banks are also used.

10.2.2   Species *versus* ecosystem conservation

Ever since humans first became interested in nature, their attention has been focused on species, which are generally easier to study than ecosystems. We have inventoried species and compiled lists of extinct species, of disappearing species, or species to be protected. Some of these species have a powerful symbolic or charismatic appeal. The panda, for example, is the emblem of an NGO (the World Wildlife Fund or WWF); the puffin is the emblem of the LPO (League for the Protection of Birds, the French affiliate of Birdlife International), and for a long time, the otter was the emblem of the Council of Europe's conservation department. Generally speaking, the 'species' approach appears firmly established in the world of nature protection and conservation. But new ideas are developing. Many feel that a policy for the conservation of biological diversity must above all strive to safeguard ecosystems, because it is illusory to protect species without at the same time protecting their natural habitats. It is for this reason that the Convention on Biological Diversity recommends the conservation of ecosystems through policies for protected areas and sustainable management. The ultimate aim is to ensure the preservation of diversity over entire ecosystems, not merely in their components. The European directive 'Habitat' and the network Natura 2000 (see Section 10.3.3) meet these priorities.

Excerpts from Article 8 of the Convention on Biological Diversity: *in situ* conservation

Each Contracting Party shall, as far as possible and as appropriate:

– establish a system of protected areas or areas where special measures need to be taken to conserve biological diversity;

– develop, where necessary, guidelines for the selection, establishment and management of protected areas or areas where special measures need to be taken to conserve biological diversity;

– regulate or manage biological resources important for the conservation of biological diversity whether within or outside protected areas, with a view to ensuring their conservation and sustainable use;

– promote the protection of ecosystems, natural habitats and the maintenance of viable populations of species in natural surroundings;

– promote environmentally sound and sustainable development in areas adjacent to protected areas with a view to furthering protection of these areas;

– rehabilitate and restore degraded ecosystems and promote the recovery of threatened species, *inter alia*, through the development and implementation of plans or other management strategies;

– prevent the introduction of, control or eradicate those alien species which threaten ecosystems, habitats or species;

– subject to its national legislation, respect, preserve and maintain knowledge, innovations and practices of indigenous and local communities embodying traditional lifestyles relevant for the conservation and sustainable use of biological diversity and promote their wider application with the approval and involvement of the holders of such knowledge, innovations and practices and encourage the equitable sharing of the benefits arising from the utilization of such knowledge, innovations and practices;

etc.

10.2.3   What are the priorities for conservation?

Is it possible to preserve the entire biosphere? This suggestion is unrealistic, because human activities necessarily involve both predator behaviour and the occupation of territory for agricultural and/or urban purposes. The logical approach, embodied in the concept of sustainable development, is to try to compromise between the economic progress necessary for the well-being of humankind, which is invasive and coercive by nature, and biological diversity, which needs free spaces. In practice, this approach relies on political decisions: which kinds of ecosystems should be protected as a priority? How should they be distributed? What criteria can help us to select which areas or species to protect?

Different propositions have been made for setting priorities:

- Protect endangered species. This is an extension of the 'species' approach that is applied in protecting e.g. the panda, gorilla, and the great African fauna in general.

- Assign priority to protecting lines of evolution that are in danger of disappearing from the face of the Earth. This could, for example, mean protecting the habitats of the Coelacanth, the last known surviving member of the Crossopterygian family. The object here is to preserve future options by protecting all the major phyla known today.

- A popular approach is to designate so-called 'hotspots,' or critical zones (cf. Chapter 2, Figure 2.6). Identifying such critical zones, where biological diversity is highly endemic and at the same time at risk, is one way of selecting priority areas for conservation. It has been calculated that by spending an average of 20 million dollars on each critical zone over the next 5 years, it would be possible to conserve a large proportion of all the world's species. Many of these critical zones functioned as refuges during the ice ages of the Pleistocene era. It is conceivable that they might also serve as sanctuaries from human activities – current and future, thus enabling flora and fauna to weather this danger. That is why they are sometimes called 'Holocene refuges.'

It must be noted that sixteen of the designated hotspots are in countries where economic conditions are difficult, prohibiting serious conservation policies, and a large part of the original environment has already been destroyed. This is the case e.g. in Sri Lanka, Madagascar, and the Atlantic forest of Brazil. By no means should the economic vulnerability of these countries be used as a pretext to turn them into expensive 'Indian reservations,' run by international authorities composed mainly of representatives from the rich countries...

10.2.4   What about the cost?

Conservation comes at a price. Theoretically, protected areas should be created in zones that are rich in biological diversity, i.e. in developing countries, who often have more pressing economic priorities than to devote their meagre resources to the conservation of species and ecosystems. At the same time, developed countries demand free access to natural resources for use in biotechnology. This causes an economic conflict between the retainers of an economically unexploited, potential treasure *vs.* industrial users, who have until now enjoyed favourable conditions, including free access to genetic resources. This North-South conflict was clearly apparent while elaborating the details of the Convention on Biological Diversity.

In theory, the Convention on Biological Diversity provides for the transfer of financial resources and technologies from developed countries to developing countries. The Global Environment Facility (GEF) controls this financial mechanism of the Convention. Since its creation in 1991, the GEF has allocated 2.2 billion dollars to activities relating to biological diversity, whilst also encouraging new investments on a national scale in all countries that signed the agreement. This sum may appear large, but it is actually risible compared with what would need to be spent. It is far removed from the annual required investment as evaluated in Rio: 125 billion dollars per year...

A global strategy for biodiversity

Under the auspices of several major NGOs for the conservation of nature, a global strategy for biodiversity has been proposed. It calls for:

- *Catalysing action through international cooperation and national planning.*
- *Establishing a national policy framework for biodiversity conservation:*
  - Reform existing public policies that invite the waste or misuse of biodiversity.

continues overleaf

A global strategy for biodiversity (*continued*)

- ○ Adopt new public policies and accounting methods that promote conservation and equitable use of biodiversity.
- ○ Reduce demand for biological resources.

- *Creating an international policy environment that supports national biodiversity conservation:*

  - ○ Integrate biodiversity conservation into international economic policy.
  - ○ Strengthen the international legal framework for conservation to complement the Convention on Biological Diversity.
  - ○ Make the development assistance process a force for biodiversity conservation.
  - ○ Increase funding for biodiversity conservation, and develop innovative, decentralized, and accountable ways to raise funds and spend them effectively.

- *Creating conditions and incentives for local biodiversity conservation:*

  - ○ Correct imbalances in the control of land and resources that cause biodiversity loss, and develop new resource management partnerships between government and local communities.
  - ○ Expand and encourage the sustainable use of products and services from the wild for local benefits.
  - ○ Ensure that those who possess local knowledge of genetic resources benefit appropriately when it is used.

- *Managing biodiversity throughout the human environment:*

  - ○ Create the institutional conditions for bioregional conservation and development.
  - ○ Support biodiversity conservation initiatives in the private sector.
  - ○ Incorporate biodiversity conservation into the management of biological resources.

- *Strengthening protected areas:*

  - ○ Identify national and international priorities for strengthening protected areas and enhancing their role in biodiversity conservation.

○ Ensure the sustainability of protected areas and their contribution to biodiversity conservation.

• *Conserving species, populations, and genetic diversity:*

○ Strengthen capacity to conserve species, populations, and genetic diversity in natural habitats.
○ Strengthen the capacity of off-site conservation facilities to conserve biodiversity, educate the public, and contribute to sustainable development.

• *Expanding human capacity to conserve biodiversity:*

○ Increase appreciation and awareness of biodiversity's values and importance.
○ Help institutions disseminate the information needed to conserve biodiversity and mobilize its benefits.
○ Promote basic and applied research on biodiversity conservation.
○ Develop human capacity for biodiversity conservation.

10.3 Protected areas

The generic term 'protected areas' is applied in connection with a number of different situations, ranging from large reserves for flora and fauna to small sites dedicated to the conservation of particular species. These may be integral reserves, excluding human intervention, or inhabited zones in which the protection of flora and fauna is ensured by involving local populations in the management of the environment and its species. At present, there are an estimated 4500 protected areas in the world, representing 3.5% of its surface above sea level.

10.3.1 National parks: nature *versus* humans

Toward the end of the 19th century, the general feeling was that conservation and exploitation of natural environments were two incompatible activities. It was therefore deemed necessary to remove entire sections of nature from human contact, which was considered to be the principle factor of disturbance. One of the first national parks in the world was created in this abiding spirit: the United States' Yellowstone Park, inaugurated in 1872. Although it is now fashionable to criticise it, this approach is not entirely void of common sense if practiced judiciously.

10.3.2   Integrating Nature and humans

But creating protected areas within inhabited zones poses social problems. In some cases, it has been necessary to resettle local populations and deny them access to zones they previously used. Under such circumstances, they are hardly motivated to respect regulations that the responsible administrations find difficult to apply for lack of sufficient means. This situation encourages poaching and sometimes engenders real social conflicts.

The example of the Galapagos National Park in Ecuador, one of the most prestigious in the world, illustrates the difficulties of reconciling the rationales of different actors: the naturalists who created the park in 1959 as a laboratory to study evolution; the inhabitants of the archipelago, whose territory is currently limited to four islands, while the population (15 600 inhabitants in 1998) is seven times what it was at the time of the park's creation; and the government of Ecuador, interested in developing a global tourist magnet. The attraction (66 000 visitors in 1999) has proved lucrative for enterprises outside the island, but has brought little benefit to the inhabitants of the park itself. The result is rapid deterioration of the natural environment due to over-frequentation, and impoverishment of the inhabitants, who have resorted to intensive fishing for exportation, causing ecological damage.

This is why planners have increasingly encouraged participation of local populations in the conception and management of protected areas towards ensuring the sustainability of such projects. In order to encourage indigenous populations to improve their management of biological diversity, it is necessary to improve their living standard and give them economic incentives for practicing effective conservation. Thus, some projects integrate conservation and development with the object of reconciling the development of new economic activities with conservation activities.

The example of the *Biosphere Reserves* is interesting in this connection, particularly as the project has an international impetus. The concept of biosphere reserves was first proposed in 1974 by UNESCO's MAB (Man and the Biosphere) programme. What distinguishes it from the traditional perception of reserves is that it tackles the joint objectives of conservation and development simultaneously.

The *Biosphere Reserve Network* currently comprises 409 reserves distributed over 94 countries. They are conceived in response to the question: how can the conservation of biological diversity and resources be reconciled with their sustainable use? The reserves represent an attempt to act on the principle that local populations constitute one of the driving forces of conservation, and that it is impossible to preserve large regions by excluding them. Local populations, management bodies, conservation movements and scientists must together search for solutions towards reconciling the conservation of biological diversity with economic development.

In principle, each reserve comprises (1) a highly protected central area, (2) a buffer or intermediary zone surrounding and adjoining the central area, where human activities such as education and ecotourism can develop, providing they comply with ecologically viable practices, (3) a transitional zone, devoted to developmental activities which may include human habitats, farming or other usages.

The Seville Conference (1995) defined the role of these biosphere reserves for the 21st century and attempted to identify their specific functions (Seville Strategy):

– a conservation function: to preserve genetic resources, species, ecosystems and landscapes, as well as cultural diversity;

– a place to gain experience in land management and regional development and to implement models for sustainable development;

– and a logistic support function: to support research projects, environmental education and training, and continuous monitoring.....

The effectiveness of conservation in protected areas varies enormously throughout the world. It must be said that many of the regions designated as protected are in reality not protected, due to lack of qualified personnel, financial resources, adequate ecological expertise, or conflicts with local customs. Consequently, they remain menaced by agricultural and urban expansion, as well as poaching.

10.3.3   Europe and biodiversity: Natura 2000

The Bern Convention of the Council of Europe on the conservation of wild fauna and natural habitats in Europe was held in 1979. Its objective was to promote cooperation among the nations of Europe in order to ensure the conservation of wild flora and fauna and their natural habitats. Special attention was focused on species (including migratory species) that are vulnerable and threatened by extinction, such as the European white stork, the crane, or the griffon vulture. The 'Bird Directive,' adopted in 1979, was the first step towards ensuring the long-term protection and management of all bird species living in the wild within the communal territory of the Council members, as well as their habitats. The Member States are responsible for safeguarding this communal patrimony of all Europeans, with particular emphasis on protecting the natural habitats of migratory birds.

The Bird Directive inspired the European Union to enact its Directive of 1992 for the conservation of natural habitats (known as the Habitat Directive). Its aim is to ensure the preservation of biological diversity by conserving natural habitats as well as wild flora and fauna, with special emphasis upon species of communal interest. It envisages the implementation of a network of protected areas called Special Areas of Conservation (SACs).

The NATURA 2000 network is an ecological network of protected areas in Europe. Comprising the relevant sites of the European Bird (1979) and Habitat (1992) Directives, its purpose is to preserve biodiversity, especially in rural and forest areas. The idea is to promote a form of management that is beneficial to the habitats of wild flora and fauna, whilst also taking economic, social and cultural constraints into consideration, as well as the regional and local particularities of each Member State. It is not the intention of the Natura 2000 network to create 'nature sanctuaries'; human activities (including hunting) are incorporated, as in the biosphere reserves.

The Member States proposed a list of sites to the European Commission, and an official list of sites of community importance was compiled in liaison with the 'Nature' theme department of the European Environmental Agency, forming the Natura 2000 network. The Member States committed themselves to establish such sites as special conservation zones and take the necessary measures to ensure their protection.

To apply the directives and implement a coherent conservation policy, it is necessary to draw up a typology of the habitats in question to enable their description and monitoring. An important source of information, the European Nature Information System (EUNIS), was implemented to harmonise the terminology and facilitate the use of data. Habitats were classified on the basis of earlier European initiatives: the CORINE Land Cover and the Classification of Palearctic Habitats of the European Council.

10.4 Sustainable use of Biological Diversity

The policy of protected areas is a stopgap solution for the short and medium term. Many people believe that the goal of biodiversity conservation must be pursued within the larger context of sustainable development. More generally, conservation should be an integral part of a global vision of land management, defining zones for agriculture, industry or urban living, as well as areas to be protected. So far, this is rarely the case.

10.4.1 Sustainable development

The concept of sustainable development is a compromise that acknowledges the validity of development but tries to reconcile the process of economic development with environmental protection. From a long-term perspective, sustainable development is development that meets the needs of the present, whilst at the same time preserving the world's 'natural' heritage for future generations.

Practical application of the concept of sustainable development is founded upon the idea that biological diversity can only be preserved in well-functioning ecosystems. Emphasis is on the need for integrated management of environments and resources. This counterbalances the narrow approach that prevailed for too long and was preoccupied with deriving short-term profits from exploiting certain resources.

The concept of sustainable development breaks with the ideology of 'development' and 'modernisation' that prevailed after the Second

Sustainable development has three dimensions:

an ecological dimension
Sustainable development asks what limits need to be placed on industrialisation in order to preserve natural resources. The object is to manage and optimise utilisation of natural capital rather than squander it.

an economic dimension
With a view towards the present and future impacts of the economy on the environment, this approach considers the choices involved e.g. in financing and improving industrial technologies, as they relate to natural resources.

a social and political dimension
Above all, sustainable development is an instrument of social cohesion and a process of political choice. Priority should always be given to equitability – between generations and among States. Any reconciliation of environment and economy must stand up to this double imperative.

World War. Sustainable development means modifying production methods and improving consumer habits, as well as changing the everyday behaviour of individuals. In theory, if not necessarily in practice, the Western model of development is no longer considered to be the unique and necessary model for social development. Different forms of development correspond to the diversity of situations and cultures in the world.

Sustainable development in agriculture is central to wise biodiversity management. In practice, it is necessarily a compromise between what is economically viable, technically possible, and ecologically acceptable. Thus, in the 1960's, cultivation of high-yield crops with intensive use of fertilisers and pesticides achieved a considerable increase in productivity; however, such progress proved detrimental to the environment and biodiversity in certain regions. From the perspective of general management of renewable resources, the goal is to conceive and implement production systems that are better integrated in their environment and maintain their ecological viability. This presupposes, among other things, that cultivation systems become more diversified and that farmers devise new technological itineraries: rotations, choice of crop varieties, cultivation methods, etc.

10.4.2   Traditional knowledge

Conservation is not just a technical problem. The erosion of biological diversity is mainly due to more intensive use of resources and changes in social behaviour and habits. On the other hand, if certain communities and cultures have managed to sustain themselves over centuries, they probably succeeded thanks to a balanced exploitation of natural resources. Traditional systems for managing living resources should be investigated to discover what social behaviour patterns made it possible to maintain a state of equilibrium between nature and societies over the course of centuries. By studying these practices, it should be possible to detect management forms that might still be applied to present situations.

Humans have in fact used a wide array of especially adapted techniques to exploit a great variety of biological resources in many different environments. Ethnobiological research is teaching us more about this overall complex of knowledge and behaviour, a cultural patrimony that is transmitted from one generation to the next. Many traditional fishing societies were organised and conducted themselves in ways that were far more effective in protecting their resources than government management systems, which often prove inoperable due to lack of adequate means.

Since the conference in Rio, the prevailing feeling has been that protection of nature and preservation of biological diversity go hand-in-hand with the rights of populations to preserve their territories and lifestyles. For a long time, this vision was anathema to many biologists and certain environmental movements, who tended to think of human populations only as disturbing or degrading factors within ecosystems. Now we know that most of the Earth's ecosystems, whether they be oases, European forests, or wetlands such as the Camargue region in Southern France, are the product of human activity.

While there are many lessons to be drawn from popular knowledge, it is equally true that economic and political developments have profoundly altered social behaviour patterns. Management systems that were well suited to certain types of constraints are no longer adequate in current economic and demographic conditions.

10.4.3   Regional planning and development

It is very popular to discuss strategies for preserving biological diversity on a regional level by creating protected areas; it must, however, be admitted that in practice, such sites tend to be selected as the opportunity arises, without a clear idea of the larger picture. Experience has shown that simply defining geographic sites for protection is not in itself an entirely satisfactory strategy.

- The environmental quality in the protected areas is strongly influenced by human activities in neighbouring regions: air pollution, changes in the water table level, deterioration of resting and feeding spots outside the protected area for migratory species, etc..

- Natural dynamics linked e.g. to climate fluctuations in protected areas may significantly alter the composition and structure of ecosystems, sometimes incurring the loss of the rare species that the reserve had been established to protect.

- The increasing fragmentation of habitats and countryside severely disturbs ecological processes and population dynamics. The maintenance or restoration of connections between different fragments to allow for functional exchanges is a major concern of conservation biology.

10.4.4   Marine environments: how can the perpetuation of fisheries be reconciled with the conservation of marine resources?

One of the major stakes of marine fisheries resides in the possibility of reconciling long-term conservation aims for species and ecosystems with the short-term demands of economically profitable fishing, be it on a small-scale or industrial basis. In fact, the history of fisheries epitomises the difficulty of applying the principles of sustainable development in practice. It highlights the question of the relationship between economic interests and biodiversity conservation, and how multidisciplinary knowledge can be integrated towards improving resource management.

Until the beginning of the 20th century, there was a tendency to believe that halieutic resources were practically inexhaustible. Scientists interpreted the evident variability of the resources as the result of environmental pressures. There was never any question of limiting catches, and industrial exploitation developed rapidly during the first half of the 20th century. After the Second World War, at the first signs that certain marine populations were overexploited (the International Convention for the Regulation of Whaling was drawn up in 1946), the concept of 'rational management of fish stocks' began to evolve. The underlying idea was to maintain balanced production by evaluating the intensity of fishing activity on the one hand and the size of the fish stocks on the other. In theory, it would then be possible to use biological expertise and models to deduce maximum levels of sustainable catch (quotas) and propose adjustment measures regulating fishing activity and/or the mesh of the fishing nets. *A priori*, this was a reasonable approach, but its limits soon became evident. In fact, many fish stocks collapsed for a number of different reasons: disregard for the quotas, rapid techno-logical development (acoustic detection, rear-trawling, the use of syn-thetic fibres to fabricate stronger nets), profit maximization, poor scientific assessment, etc..

On the global scale, the example of the fisheries is a good illustration for the principle that freely accessible biological resources tend to be overexploited. It is difficult to reconcile the discussion of long-term resource management with the economic constraints that lead the owners of fishing boats to strive for maximum profits in minimum time. Rather than fish more, they should fish better and avoid overexploitation by restricting catches. But in reality, economic competition induces irrational behaviour patterns in the absence of binding regulatory measures.

The growing scarcity of marine resources has elicited different re-sponses. In the early 1980's, coastal countries appropriated all resources contained within their exclusive economic range up to 200 marine miles distance from the coast. This move was intended to preserve resources for the use of the coastal country, which might also charge foreign fishing fleets for access. In principle, this measure countered the overexploitation of fish stocks. Another reaction designed to keeping certain fishing fleets afloat was to prospect new geographic zones (Southern seas and the periphery of Antarctica, for example) and new resources (deep-sea species...).

Where do we stand now? The annual global catch from the oceans is currently estimated at around 100 million tons and will probably peak at this level, since, according to the report on the global state of marine resources published by the FAO (www.fao.org/fi/default.asp), two-thirds of the stocks are overexploited. This situation is all the more worrying, because research has shown that most marine fish stocks do not regenerate quickly when fishing activity is ceased. Certain natural phenomena such as El Niño can also severely impact the abundance and the availability of exploitable resources, especially where the pressure from fishing is strong. Since the time it takes for them to regenerate is incompatible with the rate at which resources are being extracted, halieutic stocks are heading towards exhaustion. The use of non-selective fishing methods (trawling, for example) also destroys great numbers of non-commercial species or disturbs benthic habitats. Massive extraction of species belonging, for the most part, to higher trophic levels, has cascade effects upon marine ecosystem functioning and causes significant modifications in marine trophic networks.

This pessimistic analysis has raised public and government awareness for the need to protect marine resources, an objective supported by a number of ecological associations. Over recent years, concepts deriving from terrestrial ecology, such as establishing reserves, publishing list of species in danger of extinction, as well as protecting and restoring habitats, have been adapted to the marine environment. Different initiatives have been undertaken:

- An increasing number of marine reserves has been set up, enabling access to resources to be regionally controlled. As in terrestrial environments, the prime object of marine reserves is to protect endangered species or ecosystems. But protected marine areas also have the function of protecting living resources. Whether they involve areas for reproduction and spawning, or habitats enabling species to avoid capture, their purpose is to preserve the conditions for regeneration of stocks. Many experiments have demonstrated the effectiveness of these measures. The European Union currently maintains 33 protected marine areas in the Mediterranean, corresponding to a total surface area of almost $5\,000\,km^2$.

- Control is exercised via the market through certification of marine products (eco-labelling), with the support of multinational companies associated with NGOs.

- On the initiative of the FAO, a code of conduct for responsible fishing was drawn up in 1995. The precautionary approach recommended by the FAO[1] for living aquatic resources is a 'mild' form of the precautionary principle. It proceeds from the simple principle that given the incertitudes of our knowledge of resources and exploitation systems, it is better to make prudent decisions and avoid unacceptable or undesirable situations. The code tends to assign greater importance to long-term considerations in contexts where priority is often given to the short-term. From the perspective of sustainable development, the point is also to reduce the cost to future generations of decisions taken by our generation.

The precautionary approach[2] recognises that good ecosystem management is indispensable to wise resource and biodiversity management. It is clearly essential to preserve the basic functions of the ecosystems, because biodiversity can only sustain itself in well-functioning ecosystems.

In 2001, the Reykjavik declaration of the FAO (www.refisheries2000.org) formulated an attempt to reconcile the interests of exploiting *vs.* conserving marine ecosystems on the basis of the precautionary approach. It recommended further research into marine ecosystem functioning, whilst at the same time urging the implementation of management measures, especially mechanisms designed to reduce excessive fishing activity. The co-viability of systems of exploitation and the natural ecosystems upon which they depend constitutes an emerging theme in which the concept of sustainability takes on a new dimension.

10.5   *Ex Situ* Conservation

Living collections are found in botanic gardens and zoos, conservatories, and public and private arboreta. They play a fundamental role in the conservation of disappearing species and in programmes for re-introducing them into the natural world. They are an essential tool for the genetic resource management of useful plants and domestic animals.

[1] Garcia, S.M., 1996. The precautionary approach to fisheries and its implications for fishery research, technology and management: an updated review. Pp 1–63 in Precautionary approach to fisheries, Part 2 – Scientific papers. FAO Fisheries Technical Paper 350/2.

[2] Garcia, S.M., 1994. The precautionary principle: its implications in capture fishery management. Ocean & Coastal Management, 22:99–125.

Convention on Biological Diversity, Article 9
– *Ex situ* Conservation

Each Contracting Party shall, as far as possible and as appropriate, and predominantly for the purpose of complementing *in situ* measures:

(a) Adopt measures for the *ex situ* conservation of components of biological diversity, preferably in the country of origin of such components;

(b) Establish and maintain facilities for *ex situ* conservation of and research on plants, animals and micro-organisms, preferably in the country of origin of genetic resources;

(c) Adopt measures for the recovery and rehabilitation of threatened species and for their reintroduction into their natural habitats under appropriate conditions;

(d) Regulate and manage collection of biological resources from natural habitats for *ex situ* conservation purposes so as not to threaten ecosystems and *in situ* populations of species, except where special temporary *ex situ* measures are required under subparagraph (c) above; etc.

10.5.1   Botanic gardens

Around 1600 botanic gardens exist around the world. The earliest collected plants for traditional pharmacology. Later, they served as acclimatisation gardens for tropical species brought back by travellers, which they tried to domesticate or use to develop new cultures for economic or decorative purposes. Botanic gardens assemble flora from conquests, commercial trade and exploratory expeditions. In recent times, their mission has changed once again. New establishments specialise in the flora and fauna of a particular environment (often a local region) with a view towards developing conservation biology and educating the public.

> The toromiro (*Sophora toromiro*) is a tree endemic to Easter Island, whence it disappeared around 1960. Several botanic gardens that had preserved seeds networked to create reproductive populations for implantation on Easter Island. Given the growth rate of the tree, it will be another hundred years before we know whether their reintroduction has been successful.

Botanic gardens have always exchanged specimens and information among themselves and with private collectors. While these activities contribute towards enlarging the range of registered plant diversity, the chosen samples tend to over-represent certain groups (orchids, cacti, carnivorous plants, bulb plants, ferns, vegetables, conifers...) to the neglect of others. On the other hand, the trade in medicinal plants, decorative plants, fruits and vegetables has developed. The scientific and legal framework for this activity poses difficulties on account of the conflicting interests involved... Nevertheless, botanic gardens have developed restricted networks based on acknowledged competencies and charters. These networks are currently developing computer data warehouses, partly accessible to general public through the Internet. Thus, around 350 institutions participate in the BGCI (Botanic Gardens Conservation International) network, compiling more than 250 000 entries on a total of approximately 30 000 species. Botanic gardens are working on refining methods for cultivating, propagating and conserving plant species and communicating this expertise to nurseries and the public. Such practices draw directly upon advances in physiological research on dormancy, cold and drought resistance, vegetative and sexual reproduction, cultivation of cells and tissues. The effects of fashion and trade add to the fascination for certain taxa, whilst more 'humble' species tend to be forgotten, even if they are in danger of extinction.

Botanic gardens play an important role in the biological diversity policy of a country. They provide the public and interested associations with a source of information and a forum for independent debate. They participate in campaigns to restore degraded environments and to reintroduce vanished local species (sensitising the public, providing specimens, assuming co-responsibility for managing indigenous environments: the flora of islands, mountains or wetlands). Together with public institutions and private enterprises, they are active partners in programmes for exploiting plant genetic resources (seed banks, data warehouses, ...).

Technological and financial investment, the need for daily maintenance and qualified personnel add up to expenses that not all economies can easily bear. The largest botanic gardens are situated in the developed countries. This reflects a flagrant imbalance in relation to the tropical regions, which are richer in natural biodiversity but economically less well endowed.

10.5.2   Zoological parks

Public and private zoological parks, as well as specialised exhibitions of live animals (often centred around spectacular species: fish, snakes, birds, insects...), have similar missions and activities to those of botanic gardens. Numbering over 2000 around the world, they are mostly organised in networks. Large data bases have been compiled, and in principle, information on approximately 250 000 live specimens, belonging to 6000 species, is freely accessible. In order to keep almost a million tetrapods and as many fish alive under artificial conditions, animal collections must meet high costs that not all economies are able to bear. They require close cooperation with veterinary science and research. Their partnership with animal friends and the general public is very strong, buoyed by considerable goodwill and enthusiasm. Interest is further stimulated by the live animal trade, which is vigorous but difficult to delineate with accuracy.

The environmental dependencies of aquatic species pose a severe constraint. Nevertheless, there is a long history of aquariums and marine stations constructed to house and exhibit creatures of the sea. These institutions enjoy a longstanding relationship with the public. Their mission encompasses information, education, conservation and research. Curiously, considering their cost, they are more evenly distributed around the planet than botanic or zoological gardens and give a fair representation of marine ecosystem diversity. Their maintenance cost is often integrated in government and regional networks for halieutic resource management or comes from basic and applied research programmes.

Marine data banks were coordinated and computerised at an early stage. They are very rich in information and much consulted. Universities use marine stations for teaching purposes; research programmes range from systematics and developmental biology to industrial pharmacology. Marine stations have the infrastructures and means to conduct ecological experiments on a real-life scale. Recently, they started to assume monitoring functions for the environment and biodiversity of littoral and coastal ecosystems that are directly subjected to human pressures (urbanisation, fragmentation, extraction, waste and pollution, etc.). They have also contributed to the follow-up and alleviation of the consequences of marine catastrophes.

10.6 Conservation Biology

Conservation biology was born in the late 1970's. Its object is to ascertain the impact of human activities upon species, communities and ecosystems and to make concrete proposals for averting ecosystem degradations. Whereas environmental protection is essentially an attempt to shield spaces and species from human actions through regulations, conservation biology adapts concepts and theories borrowed from ecology and develops them further to propose appropriate methodologies and implement concrete activities for nature conservation. Like other 'crisis' disciplines combining theory and practice at the cross roads between science and management, conservation biology puts the emphasis on action. Its proponents are working under pressure, since endangered species and habitats may soon disappear, unless effective measures are taken.

Conservation biology finds application in the implementation of the Convention on Biological Diversity. It tackles the task of meshing possible options for conservation with theoretical advances in biology, genetics and biogeography, whilst also taking ethological, physiological and economic inputs into account.

If at first conservation biology focused on symbolic or charismatic species, it soon became evident that the problem of conserving habitats

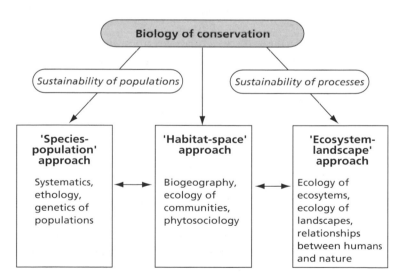

Figure 10.1 The three main approaches to conservation biology (according to Barnaud, 1998)

at the local and global level was becoming at least as important as knowing about the biology of species. Restoring and rehabilitating habitats and reintroducing species are increasingly important issues in reconstituting biological diversity. They demand both *ex situ* and at the same time *in situ* approaches. Methods for implementing the latter are improving rapidly.

10.6.1   The fragmentation of habitats

One of the favourite themes of conservation biology concerns the fragmentation of natural habitats due to human activities and its consequences for biodiversity. According to the dynamic equilibrium theory of island biogeography, the number of species present in an ecosystem is a function of the surface of the ecosystem. Thus, in principle, a reduction in surface area fosters the extinction of certain species. Conservation biologists are called upon to answer more precise questions about the size and form of natural reserves:

- What is the minimum size for a reserve for the protection of a particular species?

- Is it preferable to create a single, large reserve, or many small ones?

- How many individuals of an endangered species is it necessary to protect in a reserve in order to prevent their extinction?

- When several reserves are created, should they be close to one another, or at a distance? Should they be isolated, or linked by corridors?

These questions provide a good opportunity to put certain ecological theories to the test. Scientists advocating the dynamic equilibrium theory of island environments believe that large-scale reserves provide a greater diversity in habitats and shelter a greater variety of species than small-scale reserves. Moreover, the populations of each species are larger, and edge effects are less significant over large areas than small areas.

Other scientists believe that, on the contrary, several small reserves make it possible to protect a greater diversity of habitats and variety of rare species over a total surface area equivalent to one large reserve. They also advance the consideration that having numerous smaller reserves lessens the risk of catastrophes such as fire, epidemics or the danger of an introduced species destroying entire populations in a reserve.

With experience, a consensus is evolving. There is a recognition that protected areas must be defined on a pragmatic basis, in accordance with precise conservation objectives, rather than on the basis of general theoretical criteria that are difficult to apply.

10.6.2 Reintroduction of species

Ex situ conservation is not only an alternative, but also a complementary approach to *in situ* conservation. *Ex situ* conservation provides reservoirs of individuals for infusing and restocking wild populations of endangered species. In reality, there are situations where habitat degradation is such that it is impossible to maintain viable *in situ* populations. The alternatives are simple: either let the species disappear forever; or try to save endangered populations by preserving them temporarily *ex situ*. They could be reintroduced into their original environment if and when the threats to the species and their habitats have disappeared. In cases where habitats have been destroyed, there is still the possibility of reintroducing the species into habitats similar to its original ones. This practice is known as translocation.

The practice of preserving endangered species in zoological parks can be evaluated on its achievements: over two centuries, they have preserved over 90% of their initial biological diversity and maintained populations of 250 to 500 individuals with a view towards reintroducing species into their natural environments. A number of zoos collaborate in programmes for breeding over 300 endangered species in captivity. The preserved populations serve as genetic reservoirs in support of the animals' survival in nature. Zoos have successfully preserved species such as the European and American bison, the Prejwalski horse, etc.

10.6.3 Restoration ecology

Engineers and specialists in the environmental protection have long attempted to 'repair' degraded environments. They have often proceeded empirically, applying their field experience to objectives defined by the criteria of ecological management of environments and species. Generally speaking, scientists neglected this domain for too long, because it was considered too technical. However, some scientists have come to realise that such activities can be of great interest if they are run as experimental approaches. Scientific interpretations of numerous ecological interventions and manipulations laid the foundation for the emergence of restoration ecology in the 1980's. Implicitly, restoration ecology believes in the opportunity of conducting rigorous ecological experiments that enable a predictive approach.

The terminology of restoration ecology is quite complex, and it is risky to attempt universally accepted definitions of the terms used. It is most important to be aware that there are different kinds of restoration, running the gamut from the reconstitution of totally devastated sites, e.g. through strip mining, to operations of a limited scope within slightly disturbed ecosystems.

- *Restoration* (*sensu stricto*) is the intentional transformation of an environment to re-establish what is considered to be its indigenous and historic ecosystem, including its original taxonomic composition as well as its pre-existent basic functions (production, autoreproduction).

- *Rehabilitation*: if the pressure exerted upon an ecosystem was too strong or long lasting, it is rendered unable to return to its former state, even if the human pressures are removed. Only massive human intervention over a limited time may be able to steer the ecosystem back onto a positive trajectory towards re-establishing its essential functions;

- *Reaffectation*: when an ecosystem has been severely transformed by humans, the solution may be to find a new use for it, rather than try to rehabilitate it. This is the case e.g. when an ecosystem is deliberately modified with the intention of favouring a particular element or function. The new state may bear no structural or functional relation to the pre-existing ecosystem, as in the case of areas under cultivation.

For some, restoration biology also includes 'creation,' i.e. creating new habitats where none formerly existed.

To cite some examples of applied restoration biology:

- reclamation of sites that have been severely damaged by exploitation and development (landfills, mines, ski slopes, gravel pits, road works, etc.). The sites of abandoned quarries cover large areas and are prime environments for 'ecological reclamation.'

- the desired return of symbolic species such as salmon is sometimes borne as a banner to arouse interest in river restoration.

Back in the days of Rousseau, Lake Annecy in the Haute Savoie region of France could be described as a 'pure mirror.' The lake was prized by anglers for its excellent fish, e.g. the arctic charr. By the 1950's, charr stocks were dwindling, and water transparency had decreased from ten metres at the beginning of the century to about five. The cause was soon identified: sewers were discharging untreated water that was highly polluted with nutrients. There were clear signs that the lake was heading for eutrophication: reduced water transparency; increased turbidity linked to the development of phytoplankton following the influx of nutrients. In 1957, the riverain communities decided to construct a collecting drain encircling the whole lake and feeding into a sewage treatment plant. The works were completed in 1972. Since then, the lake water has slowly improved in quality. In 1993, its transparency reached 12 metres, the same as at the beginning of the century.

10.7 The Preventive Approach: Ecosystem Health Checks

Managers need tools to help them promote acceptable ecological practices. They turn to ecologists to enquire how to evaluate the ecological state of ecosystems. The concept of ecosystem health is an integrated approach that puts the emphasis on the 'quality' of biological diversity and provides rules of conduct for restoring damaged ecosystems. Biological diversity is an essential element of such bills of health, because it reflects the consequences of physical, chemical or biological modifications wrought upon ecosystem functioning. Thus, biological diversity may be regarded as mediating between ecological systems and social systems.

10.7.1   Ecosystem health and/or integrity

Extending the idea of health to ecosystems is justified by evidence that
ecosystem dysfunctioning often results from the impact of human activities. A system in good health is defined as a system that is able to
maintain its organisational and functional autonomy over time.

We need criteria and methods to identify dysfunctions, to
evaluate their causes and to suggest possible solutions. A gauge for
the state of health of ecosystems is provided in part by biotic (cf.
below), physical or chemical indicators. But it is also necessary to consider the expectations of the society, which are linked to its value
system and perceptions. It is generally accepted that the expression
'ecosystem health' is usually employed subjectively to describe the
desirable state of a system as defined by a group of stakeholders. E.g.
the main criterion could be the ability to catch certain kinds of fish or
to observe certain kinds of birds. Perceptions can also differ according
to type of ecosystem and period. Altogether, the notion of health is
relative.

The Canadians have developed a related concept called 'ecosystem
integrity'. The biotic integrity of ecosystems can be defined as the capacity
of an environment to maintain a balanced and well adapted community
of organisms with a species composition, diversity and functional organisation comparable to those of local natural habitats (or at any rate, the
least disturbed habitats). The concept of integrity, like that of
health, relates to social values. In the face of disturbances, the ecosystem
must preserve its capacity to react and evolve towards a final stage that is
normal or 'good' for that particular ecosystem. In systems that have been
modified by human intervention, the question of integrity takes the form:
what kind of 'garden' do we want? It is difficult to anticipate the ecosystems of the future, but it is possible to have some idea of what would be
desirable.

Concepts of health or integrity are imbued with ethical and moral
implications. There are ecosystem states that are 'normal' and others
that are 'abnormal,' also described as dysfunctional. This normative
approach accords with the notions of variability and heterogeneity that
are currently evolving in ecology. For the moment, these concepts as
such are accepted by users and help orientate public policies, in the
absence of other, more heuristic methods.

10.7.2   Bio-indicators

It is impossible to measure all the components of an ecosystem to ascertain its state of health. Consequently, it is necessary to determine which data are the most relevant. The aim of bio-indicator research is to provide the tools for characterising the ecological evolution of an ecosystem's condition over time. In particular, bio-indicators serve to determine whether environmental conditions are deteriorating under the effect of predictable impacts. They are also useful for verifying or refuting the benefit deriving from regulatory measures and adjusting interventions that are not achieving their objectives.

Bio-indicators are not simply data about ecosystems. They should also:

- convey relevant information in a condensed form and strive to represent complex phenomena in a simplified manner;

- provide a means of communication, in particular between data compilers and users.

Following are some of the different biological variables currently used as indicators.

- At the individual level, there is a distinction between *biochemical indicators* (enzymatic modifications, carcinogenesis), *physiological indicators* (growth rates, fecundity rates, diseases), and *behavioural indicators*. Levels of bioaccumulation in organisms are also used as indicators for degrees of exposure. Such organisms are sometimes called 'sentinel' organisms.

- At the species level, scientists have identified *indicator species* that have special habitat requirements in terms of a set of physical and chemical characteristics. The presence or absence of such a species, morphological modifications, or changes in its behaviour enable assessment of the degree to which its environment is within the range of its optimal needs.

- At the population level, the focus is mainly on *demographic indicators* (age and size structures, birth and death rates, gender ratio).

- At the ecosystem level, it is possible to examine community structures (species richness, abundance, biomass, structural indicators), processes (primary production, secondary production, nutrient cycles), interdependencies (trophic levels, food chains), or landscapes (heterogeneity, fragmentation). These aspects are called *ecological indicators*. In the case of complex, multispecific ecological indicators, it may be useful to calculate *biotic indices* by using either a census or an arbitrary system of notation.

Sentinel organisms

Certain organisms may act as sentinels for changes in the environment either because they react to low rates of pollution, e.g. by growing scarcer or pullulating, or else because they have a special aptitude for accumulating the contaminants in their tissues. Lichens are a good example of organisms with a great capacity for accumulating the pollutants contained in rain water, to the point that, in some regions where the nuclear fallout from atomic explosions concentrates, they are becoming a menace to the organisms that feed on them. In aquatic environments, bryophytes (foam) acquire the local level of water contamination over the long term. On account of their capacity for accumulating metallic pollutants and radioactive elements, they are good bio-indicators for detecting pollution.

10.8   Disturbances: Allies of Conservation?

Whilst conservation generally invokes the protection of natural environments under threat, it is now accepted that severe disturbances are sometimes necessary to maintain ecosystem biological diversity over the long term.

10.8.1   Fire

Fire is often perceived as a destructive element related to human activities. Entire ecosystems have been erased by fire. But over the course of the ages, many species and ecosystems have adapted to the fires that ravage the face of the Earth on a more or less periodic basis. Fire is not a human invention; it is a natural component of many ecosystems. Lightening and volcanic eruptions are two of its primary causes. In the

Yellowstone Park region, for example, lightening ignited at least 369 fires between 1972 and 1987.

In 1988, an unusually fierce fire devastated around one-third of Yellowstone National Park in the USA. Some ecologists were quick to proclaim this an ecological disaster that would alter the region's landscape forever. Ten years later, the actual outcome is much more subtle. Shrubby and herbaceous vegetation has once again taken hold, and the present landscape is just as heterogeneous and diversified as before the fire of 1988. All things being equal, the natural biological systems have regenerated rapidly.

For forests that are actually dependent upon fire, including many boreal, Mediterranean or dry tropical forests, fire is a foreseeable event that is perquisite to good ecosystem functioning. Fire increases the variety of the forest in terms of its species composition, size and demographics. It can have the following beneficial effects:

- By removing trees, fire opens the forest canopy and enables the emergence of pioneer species. In some ecosystems, forest fires also have function of reducing the overall surface of the forest, allowing grasslands and steppes to develop.

- Fire favours the germination of species (which shepherds sometimes call 'the daughters of fire') whose germination is triggered when flames burst their fruit and liberate the seeds. Scientists have observed that the Alep pine emits large quantities of fertile seeds in the weeks following a fire.

- Fire facilitates nutrient leaching in soils, especially when plants have coarse, slowly decomposing leaves.

Fire is also an intrinsic constituent of the savannas that cover almost one-fifth of the ground above sea level. Savannas have a number of characteristics that are favourable to fire: dry climate periods alternating with humid episodes; dry storms with lightening; dry wood and grass that ignites easily, etc..

From the early Tertiary until the mid-Miocene period, the Earth's surface was covered by dense forest. From 10 to 12 million years ago, savannas and herbaceous zones started to spread throughout the Tropics; five million years ago, they covered vast expanses. The grasslands of today appeared around 2 million years ago, during the time of

Homo erectus. Dryer and colder climate conditions are usually cited to explain this phenomenon. Grasslands were probably maintained by fire as well as by herds of large animals, whose grazing also contributed towards clearing forested areas.

'Accidental' fires may also occur in the wrong places, in the wrong season, and with the wrong degree of intensity. They alter the vegetation and eliminate species that are not well adapted to survive fires. On the other hand, changes in the fire regime and related disturbances to landscapes and forests may constitute a threat to these environments. When fires are suppressed, trees invade open systems, and species adapted to the passage of fire disappear from the clearings.

Today, with urban sprawl encroaching upon environments, it is necessary to take protective measures against fire. But injunctions against burning lead to an accumulation of waste, which in turn renders the plant cover more flammable and causes more intensive fires. The question is how to reintroduce fire as a means of environmental protection and conservation in zones where it plays an ecological role.

10.8.2   Cyclones and storms

Storms are among the most violent and least predicable of meteorological events. Their consequences for forests are known: by causing windfalls, storms contribute towards reinitiating the process of succession, fostering the maintenance of large species richness by reinstalling pioneer species. Recolonisation occurs through sprouts, the germination of seeds contained in the soil, or an influx of seeds from neighbouring populations.

Windfalls typically create microhabitats, as in dead wood, that provide many animal and plant species with materials for nutrition, nesting or shelter. The decomposition of upright dead wood by mushrooms or insects forms cavities that are frequented by many different animals: squirrels, martens, dormice, ferrets, common genets, etc., who use them as a daytime shelter, while a number of cavernicolous birds use them for reproduction: woodpeckers, nuthatches, tits, treecreepers, tawny owls, stock pigeons, hoopoes, etc.

Hurricanes and cyclones contribute to the maintenance of forest diversity. Ten years after Hurricane Joan swept through Nicaragua, the devastated forest zones have two to three times more tree species than the zones that remained untouched. The proliferation of new species appears to be greatest where the initial destruction was most severe.

10.9   International Conventions

Recent decades have seen the ratification of several specialised conventions intended to ensure better protection of nature. To name a few:

- The *Ramsar Convention*, named after the resort in Iran where the international treaty was signed in 1971, has the object of conserving wetlands of international significance, with special emphasis on those used as habitats by waterfowl. The signatory countries pledge to create wetland reserves and place at least one of these on the global list of important wetlands.

- The *Washington Convention* or *Convention on International Trade in Endangered Species of Wild Fauna and Flora* (CITES) was signed in 1973. Its objective is to regulate international trade in animals and plants, both dead and alive, as well as all their parts or recognisable derivatives. The Convention prohibits almost all trade in endangered species.

- The *Convention on Biological Diversity*, which has been in effect since 1994, is the first global instrument to take into account all aspects of biological diversity: genetic resources, species and ecosystems. The Convention advances several important principles:

  – *sustainable conservation and management of biodiversity*. Biodiversity conservation should prioritise *in situ* conservation of ecosystems and natural habitats, either in protected areas where special measures can be taken to preserve biological diversity, or by sustainable management of environments and exploited resources. Emphasis should also be placed on maintaining and preserving traditional knowledge and customs of indigenous communities that are relevant to the sustainable use of biological diversity. It may also be necessary to implement measures for the *ex situ* conservation of certain components of biological diversity, preferably in their countries of origin.

  – *principle of sovereignty over biological resources*. The Convention reiterates that States have sovereign rights over their natural resources and that conditions for access to genetic resources are determined by national legislation. Each State is responsible for preserving the biological diversity on its territory. Each State must take the necessary measures to ensure that activities undertaken within the confines of its jurisdiction do not cause environmental damage to other States. The Convention explicitly acknowledges economic and social development and the eradication of poverty as priorities of developing countries within the context of sustainable development.

– *access to and transfer of technology*. The Convention advances the principle that the results of industrial research should be shared on a fair and equitable basis. Benefits resulting from products developed on the basis of genetic resources provided by one of the Contracting Parties shall be the object of negotiated agreements. The Signatories agree to facilitate the transfer of and access to technologies to promote the conservation and sustainable use of biological diversity.

• The *Convention on Desertification*, which did not come into effect until 1996, aims to put an end to a worrying phenomenon: around one quarter of the ground above sea-level is in the process of desertification, i.e. undergoing soil degradation and progressively depriving the Earth of its agricultural and grazing potential, whilst simultaneously destroying biological diversity. This is not a case of traditional deserts expanding, but rather of arable land gradually being transformed into sterile space as a result of human pressures. The Convention on Desertification promotes countermeasures such as soil regeneration, improved crop yields, tree plantations, etc. Unfortunately, they suffer from a crucial lack of funding.

10.10   From Theory to Practice: Some Examples for The Implementation of the Principles of Conservation and Sustainable Development, and The Difficulties Encountered

It is not enough to formalise rational, general principles within the framework of a convention on biological diversity and sustainable development; the next step is to take concrete actions to promote biological diversity conservation. Given the diversity of existing situations, it is illusory to imagine that there is one, universally applicable theory of conservation. In reality, pragmatic approaches have usually prevailed. Oriented along generalised guidelines for the implementation of international conventions by their ratifying States, such approaches are founded upon nationally or regionally defined policies and also take the form of local activities.

National policies

Each country develops its own policy on biodiversity. The great variety of measures adopted around the world reflects very different political and cultural situations and the different influences exerted by history, power structures, and the economic realities of rich and poor countries.

With variations according to country, applications of the Convention on Biological Diversity address biological, ecological and environmental aspects of the issue. At the national level, policies on matters of genetic resources and agriculture take their cue from the FAO propositions, while legislation governing intellectual property and commerce follows the recommendations of the WIPO and WTO. On the other hand, the way in which national policies are structured and coordinated also bears the stamp of national cultural identity. It is a delicate task to reconcile the different and sometimes contradictory demands involved in such tasks as inventorying national patrimony, regulating the protection of natural sites, defining new rights, changing the rules of land use and property, etc., not to mention the pursuit of development, which also remains a priority. In Finland or France, the sense of cohesion is traditionally strong. In many European countries, as in Canada and the

Citing the Convention on Biological Diversity and the concept of national sovereignty, Thailand has enacted national legislation establishing new rules for the protection of and access to genetic resources, biological diversity, and traditional medicinal plants. The exact wording of this national law was still in deliberation when the American government demanded that it be outlawed by the World Trade Organisation. The USA protested against what it saw as protectionism to the detriment of medical research and claimed that national control over genetic resources constituted an impediment to free trade.... We are still far from sharing the benefits of biodiversity in a just and equitable manner ...

USA, the responsible administrations still leave considerable leeway for private as well as public action. In many developing countries, a coherent conservation policy has yet to be developed.

10.10.1   Implementing international conventions

Ten years on, perspectives have evolved considerably compared to the objectives advanced and debated at the Rio Summit of 1992. Global change and its consequences for the environment are now discussed by independent forums (e.g. the Kyoto conference on climate change) and addressed in special negotiations. In matters of biodiversity, the opposition between the countries that ratified the Convention on the one hand and the United States of America on the other has become more subtle. Many countries now see the Convention as a very general framework and believe that State sovereignty, national policy and legislation should be respected. Countries like Australia and Brazil are advocates of this position, which is also agreeable to the USA and the European Union. The debate continued at the global summit of Johannesburg in 2002.

Biodiversity, once recognised as a major element of global change and in urgent need of protection, has lost importance relative to debates and specific negotiations over genetic resources, biosecurity, intellectual property rights, etc. Questions of access to genetic resources and benefit-sharing were not new in 1992 and formed the subject of debates, negotiations and agreements (cf. Chapter 9) at the time. They are integrated into and underlie the Convention on Biological Diversity, which provided the relevant directives and philosophical guidelines. Their practical implementation follows the stipulations of the International Treaty on Plant Genetic Resources for Food and Agriculture, as negotiated by the FAO. Intellectual property rights pertaining to benefit-sharing fall within the domain of the World Organisation for Intellectual Property, and these questions are handled within the framework and under the indirect authority of the World Trade Organisation.

Turning to the issue of biosecurity, the stakes raised by the massive development of Genetically Modified Organisms in the agro-alimentary industry have engendered fierce debates and power struggles. The economic aspects of biosecurity have been analysed and are regulated by the measures adopted in the Cartagena Protocol, whilst the *Codex alimentarius* examines the impact of GMO on human health. However the acceptability of GMO and measures restricting their use are not decided by protocols alone. They are determined by men and women who are influenced by their ideas and cultures, their tribulations and pleasures, liberties and dependencies. In this domain, recourse to the precautionary principle has been of little assistance, save to show that it is not a universally applicable paradigm.

The pragmatic approach to diverse situations has resulted in specific negotiations and fragmented the field of biodiversity. This has proven extremely favourable to expressions of self-interested positions and power struggles. Certain countries (Brazil, the USA, the European Union) stand by their national sovereignty and refuse to consider any actors other than the State; others (Switzerland, Australia) are in favour of voluntary measures implemented by private actors; and finally, some countries subordinate their cooperation to political imperatives (framework law on biodiversity of the Organisation of African Unity, patent rejections by Ethiopia) or to development goals (recognition of national rights, knowledge and know-how for India and South Africa, transfers of technology in Colombia).

10.10.2 Local and human stakes in biodiversity: transcending the contradictions?

The stakes involved in biodiversity are such that at present, there is no true guarantee that it will be conserved for future generations. Large-scale measures on a national or international level are not the whole solution. It is well understood that conflicts of interest are inevitable in the short term, affecting the results of all negotiations, and that fair benefit-sharing is still a distant goal. There is no doubt that to ensure the conservation of biodiversity, compromises must be found. But it is urgent that we ask ourselves whether the present situation, whose legal regulations and institutional frameworks partly reflect our immediate interests, is at all compatible with the long-term objectives of conservation.

Social science research has focused on this question and proposed new practices. Some of these have been tested in the real world, and although they have been unable to break the mould of currently predominant practices, they nevertheless offer interesting perspectives to explore in the future. Some examples by way of illustration (Babin and Antona, 2001):

10.10.2.1 Two medicinal plants from Madagscar

Prunus africana is a tropical secondary forest tree valued for various uses, in particular as the source of an active ingredient contained in its bark that is used to control prostate hyperplasy. Thus, the species is exploited for both local and international purposes. The pharmaceutical market has increasing demand for the extract. Wise exploitation practices recommend removing only part of the bark from well-developed trees, without causing the plants to die or preventing them from reproducing. On the other hand, it is more profitable in the short term simply to chop down a tree and use all its bark. Non-sustainable exploitation practices have had the upper hand, and the species is now registered on the list of species that risk extinction within the next ten years. And yet, a scenario for its conservation can be envisaged. The idea is to draw up temporary contracts between the companies and the local communities for conducting a policy of resource conservation and equitable benefit-sharing. In principle, an agreement can be reached, but it is difficult to implement in practice, because it entails two fundamental challenges to social practices: the transfer of the responsibility for management of the forest from the State to the local communities; the intervention of an authorised mediator empowered to regulate relations between the strong (the State and the companies) and the weak (the local communities).

Centella asiatica is an herbaceous plant that comes from Asia. It is perennial, rampant, multiplies primarily by vegetative reproduction and thrives in humid or disturbed environments. Its rosettes are valued by traditional medicine for their cicatrising properties, as well as by the pharmaceutical and cosmetic industries, which extract active ingredients. As opposed to the plum tree discussed above, this species is not endangered. However, it is exploited without any respect for the rules of fair and equitable benefit-sharing: once harvested, the raw product is exported even though at least a portion of the transformation process could be accomplished in Madagascar and generate revenue. While this power game, which is contrary to the principles of fair benefit-sharing, is quite satisfactory for the foreign pharmaceutical companies, a contractual agreement more beneficial to the country could be envisaged under several conditions: if Madagascar were to enact a patent law, if the government were to delegate power to national enterprises for the development of specialised processing, and if contracts with foreign companies were negotiated and periodically revised.

These two examples illustrate the difficulty of implementing the good resolutions made at the international level and putting them into practice at the local level; at the same time, they highlight the importance of taking into account how societies function in order to ensure biodiversity conservation. The proposed scenarios are possible outcomes, but others are equally imaginable.

10.10.2.2   Towards new social contracts?

Long-term collective interests usually carry little weight in the face of national or private strategies. Among the possible ways of involving the different social partners in biodiversity management, the concept of Multiple Interest Accommodation reverses customary logic. Instead of proceeding on the basis of existing power structures, where contradictory interests tend to block the process of evolution towards conservation and equitability, negotiations among partners are focused upon long-term objectives. In such negotiations over the distant future (i.e. beyond a generation), the role of the mediator consists in giving voice to the partners who are not immediately present (silent partners), to adjust the existing imbalance of power, and to provide for periodic revision.

New approaches, introducing long-term patrimonial mediation through partners who are not direct stakeholders and seeking to reconcile multiple interests on a contractual basis, are exploring ways that may be better suited to deal with the contradictions of human societies. The future will show whether they prove effective.

Since the 1980's, there has been a conflict of interests between agricultural development in the Zambeze Valley and wild fauna conservation and management in Zimbabwe. The State has devised a local management programme for wild fauna (the CAMPFIRE programme) that entails redistributing a proportion of hunting profits to the villages. The process identifies two administrative levels: the district council, where goals for protecting and managing wild fauna are negotiated; and the villages, where questions of land and use and territory are settled. Passing incentives for conservation management down to local communities, the structure of CAMPFIRE provides for decentralised decision-making and the devolution of resource management responsibilities. The devolution of power has been quite successful in shifting attitudes away from dependency upon central institutions towards self-reliance and self-sufficiency. In the process, local institutions have also been strengthened in the realms of project management and accountability. However, whilst community participation has been effectively implemented in some areas, there has been much less success in others. The problem has usually had to do with the nature and effectiveness of the devolution process. In some cases, devolution has progressed no farther than the district councils, leaving local communities frustrated and powerless. This has led to misunderstandings and sometimes to hostility towards CAMPFIRE, accompanied by an increased mistrust of district councils, inadequate environmental supervision, and intolerance towards wildlife perceived as causing damage without providing any local benefit. In environmental terms, the overall result has been negative, marred by continued illegal poaching and further encroachment into wildlife areas. The problems of participatory conservation management have been more acute in edge zones that have fewer elephants or other wildlife valued by tourists and hunters and thus attract less investments.

In South Africa, the Communal Property Associations Act has set up a new authority responsible for communal property management. Its goal is to encourage paper companies and village authorities to form contractual partnerships. The reality is not so positive, since in the absence of legitimate, strong mediators, the imbalance of power still impedes any real progress towards more conscientious conservation and more equitable benefit-sharing.

Once they enter the domain of biodiversity management, the markets for use-rights and intellectual property rights will have a major impact in developing countries. Depending upon how they are regulated, these markets can either strengthen rural communities or else, if these rights are cornered by those parties empowered by the state, tradition or wealth, marginalize them even further. Given the present situation, the latter alternative appears more likely. The future of biodiversity is far from assured, despite the current international dialogue.

A Conclusion of Sorts

The concept of biodiversity has taken hold at the crossroads of natural sciences and social sciences. The natural sciences have been marginalized for some time and are striving to regain public interest; the social sciences are discovering the complexity, but also the richness, of the relationship between humankind and nature. Both sciences approach biodiversity as a field of application for the new relationships that are developing between humans and nature, raising new questions and concerns regarding the living world.

Biodiversity is a banner rallying all those concerned about the possible consequences of the general trend towards artificiality *versus* nature. The concept feeds upon the general feeling that human activities may endanger the future of humanity. The old, fundamentally utilitarian view of nature is being supplanted by an ethical code based upon respect for life. Prompted by different reasons, but orientated toward the same goal, scientists, nations and non-governmental organizations have each developed their own arguments and are beginning to implement various strategies designed to check the erosion of the living world.

New fields of research are emerging. The life sciences are seeking to reconcile genetics and ecology to improve our understanding of environmental impacts upon genome expressions and evolutionary mechanisms. We are moreover rediscovering that biodiversity is part of our daily life, that it may represent a considerable economic stake, and that legal experts are called upon to design effective laws for the protection of nature.

Biodiversity has become a social issue. It appeals to new moral values that question the priorities of economic models of development. A certain promise lies in the new, friendly relationship with nature that appears to be evolving in the West. Decision-makers and producers are under pressure to change their relationship with natural science specialists.

Biodiversity   Christian Lévêque and Jean-Claude Mounolou
© 2004 John Wiley & Sons, Ltd   ISBN 0 470 84956 8 (Hbk)   ISBN 0 470 84957 6 (pbk)

Scientists are no longer occupied simply with writing the necrology of species, they no longer stand by as helpless observers when major ecological disasters occur; rather, they are called upon to help degraded environments recover their biological integrity, their functions and ecological services. One instance of this role, the reinstallation of the salmon, has advanced to a symbol and turned into a qualitative standard for the European river ecosystem.

Preservation of the biodiversity that is our heritage requires local management by the populations immediately concerned. Centralized management forms, advocated by Western societies in accordance with their own perceptions of nature and moral values, have no universal validity. Conscious that it is in the nature of international law to lag behind events and that considerable economic interests are involved, it is legitimate to ask what the real implications of potential protective measures are. The shape of the future will necessarily depend upon the ways in which societies and scientists are able to make themselves heard by the policy-makers of today.

On the Web . . .

Information network

List of biodiversity information networks
www.bdt.org.br/bin21/

The World Biodiversity Information Network
http://www.conabio.gob.mx/remib_ingles/doctos/remib_ing.html

Biodiversity Resources in Belgium
http://BETULA.BR.FGOV.BE/BIODIV/

National Biological Information Structure (USA)
http://www.nbii.gov/issues/biodiversity/ecological.html

Inventories and data bases

Global

http://nature.eionet.eu.int/

GBIF www.gbif.org

Marine species : http://erms.biol.soton.ac.uk

Species 2000 : www.species2000.org

FAO-Database on introductions of aquatic species
http://www.fao.org/waicent/faoinfo/fishery/statist/fisoft/dias/
mainpage.htm

Biodiversity   Christian Lévêque and Jean-Claude Mounolou
© 2004 John Wiley & Sons, Ltd   ISBN 0 470 84956 8 (Hbk)   ISBN 0 470 84957 6 (pbk)

The World Biodiversity Database
http://www.eti.uva.nl/database/wbd.html

MAB species databases
http://www.ice.ucdavis.edu/mab/index.html

By group

Vertebrates
Birds and mammals of Latin America
http://www.natureserve.org

World Turtle Database
http://emys.geo.orst.edu/main_pages/intro.html

FishBase - a global information system on fishes
http://www.fishbase.org/search.html

Mammal species of the world
http://nmnhwww.si.edu/msw

Reptile database (EMBL)
http://www.embl-heidelberg.de/~uetz/livingreptiles.html

Amphibian Species of the World
http://research.amnh.org/herpetology/amphibia/?genus
=Ansonia&species=albomaculata

AmphibiaWeb
http://elib.cs.berkeley.edu/aw

Invertebrates
Australian Aquatic Invertebrates
http://www.lucidcentral.com/keys/lwrrdc/public/Aquatics/main.htm
marine mollusks : www.mnhn.fr/base/malaco.html
marine invertebrates : http://erms.biol.soton.ac.uk

Collections & Regional Inventories

Biodiversity and Biological Collections Web Server
http://biodiversity.uno.edu

Canada's Aquatic Environments
http://www.aquatic.uoguelph.ca

Hawaii Biological Survey Databases
http://hbs.bishopmuseum.org/hbsdb.html

Illinois Natural History Survey
http://www.inhs.uiuc.edu/cbd/collections

Muséum National d'Histoire Naturelle
http://www.mnhn.fr/base

Florida Museum of Natural History
http://www.flmnh.ufl.edu/databases/

Alaska-Nearctica
http://www.nearctica.com/syst/author/authmain.htm

New Zealand freshwater organisms
http://www.niwa.cri.nz/rc/prog/freshbiodiversity/tools

Resource Guides

The Virtual Library of Ecology and Biodiversity
http://conbio.net/VL/browse/

BIOSIS Resource guide by animal group
http://www.biosis.org/zrdocs/zoolinfo/gp_index.htm

Links on biodiversity
http://www.geocities.com/RainForest/Vines/8695

Web sites on aquatic entomology
http://www.ippc.orst.edu/dir/gateway/entomology/aquatic.html

The World Species Lists (list of checklists published on the web)
http://species.enviroweb.org

Entomology checklists
http://www.ent.iastate.edu/list/checklists.html

http://www.unep.org/vitalwater/resources.htm

Systematics

The Tree of Life
http://tolweb.org/tree?group=Life&contgroup=

Animal Diversity Web (references and systematics)
http://animaldiversity.ummz.umich.edu/

Reports & Maps

Climate and freshwater biodiversity
http://www.pewclimate.org/press_room/sub_press_room/2002_releases/pr_aquatic.cfm

UNEP-Virtual Water Graphics
http://www.unep.org/vitalwater/resources.htm

Electronics books available on line:
http://www.sil.si.edu/Subject-Guide/nhebookls.htm

Conventions

CDB : www.biodiv.org

CITES : www.cites.org

Ramsar : www.ramsar.org

Convention sur les espèces migratrices : www.wcmc.org.uk/cms

Rio Earth Summit:
http://www.unep.org/Documents/Default.asp?DocumentID=78&ArticleID=1163

Habitats Directive:
http://europa.eu.int/comm/environment/nature/habdir.htm

The World Commission on Dams:
http://www.dams.org/

Human health (emerging diseases)

Society for Environmental Education - Biodiversity and Human Health
http://www.ecology.org/biod

Conservation

World Resources Institute - Species Extinctions
www.wri.org/wri/biodiv/extinct.html

The World Conservation Union
www.iucn.org

http://www.darwinfoundation.org/

Conservation International - Biodiversity Hotspots
http://www.biodiversityhotspots.org/xp/Hotspots

WWF Living Waters Programme
http://www.panda.org/about_wwf/what_we_do/freshwater/

European Centre for Nature Conservation
http://www.ecnc.nl/doc/servers/biodiver.html#habit

The Nature Conservancy - Freshwater Biodiversity Conservation
http://www.freshwaters.org/info/specific.shtml

NatureServe
http://www.natureserve.org/

The World Bank and Biodiversity
http://lnweb18.worldbank.org/ESSD/envext.nsf/48ByDocName/
Biodiversity

References and Further Reading

References

Barnaud, G., 1997. *Conservation des zones humides: concepts et méthodes appliquées à leur caractérisation.* Thèse, Université de Rennes 1.

Clements, F.E., 1916. *Plant Succession. an analysis of the development of vegetation.* Publication no. 242. Washington, DC: Carnegie Institution of Washington.

Constanza, R., d'Arge, R., de Groot, R., Farber, S., Grasso, M., Hannon, B., Limburg, K., Naeem, S., O'Neill, R., Paruelo, J., Raskin, R.G., Suffon, P. and ven en Belty, M., 1997. The value of the world's ecosystem services and natural capital. *Nature* **387**, 253–260.

Cornell, J.H., 1978. Diversity in tropical rainforests and coral reefs. *Science* **99**, 1302–1310.

Cowx, I.G. (Ed.), 1997. *Stocking and Introduction of Fish.* Oxford: Blackwell Science, Fisshing New Books.

di Castri, F. and Younes, T. (Eds), 1996. *Biodiversity, Science and Development. Towards a New Partnership.* Wallingford: CAB International, p. 2.

Foucault, A., 1993. *Climat. Histoire et avenir du milieu terrestre.* Fayard: Le Temps des Sciences.

Holling, C.S., 1986. The resilience of terrestrial ecosystems: local surprise and global change. In: Williams C.C., Munn R.E (Eds), *Sustainable Development of the Biosphere.* IISASA. Cambridge: Cambridge University Press, pp. 289–317.

Hubbell, S.P., 2001. *The Unified Neutral Theory of Biodiversity and Biogeography.* Princeton, NJ: Princeton University Press.

Hugueny, B. and Paugy, D., 1995. Unsaturated fish communities in African rivers. *Am. Nat.* **146**,162–169.

Huston, M.A., 1979. A general hypothesis of species diversity. *Am. Nat.* **113**, 81–101.

Biodiversity Christian Lévêque and Jean-Claude Mounolou
© 2004 John Wiley & Sons, Ltd   ISBN 0 470 84956 8 (Hbk)   ISBN 0 470 84957 6 (pbk)

Huston, M.A., 1997. Hidden treatments in ecological experiments: re-evaluating the ecosystem function of biodiversity. *Oecologia* **110**, 449–460.

Hutchinson, G.E., 1957. Concluding remarks. *Cold Spring Harbor Symp. Quant. Biol.* **22**, 415–427.

Hutchinson, G.E., 1961. The paradox of the plankton. *Am. Nat.* **95**, 137–147.

Jaeger, J.J., 1996. *Les mondes fossiles.* Sciences. Paris: Editions Odile Jacob.

Kropotkin, P.E., 1988. *Mutual Aid: A Factor of Evolution.* Black Rose Books.

Laubier, L., 1991. Les marées noires. Conséquences à long terme. *La Recherche* **233**, 22.

Lubchenko, J., *et al.*, 1991. The sustainable biosphere initiative: an ecological research agenda. *Ecology* **72**(2), 371–412.

MacArthur, R.H. and Wilson, E.O., 1963. An equilibrium theory of insular zoogeography. *Evolution* **17**, 373–387.

MacArthur, R.H. and Wilson, E.O., 1967. *The Theory of Island Biogeography.* Princeton, NJ: Princeton University Press.

Macfadyen, A., 1957. *Animal Ecology – Aims and Methods*, 1st edition. London: Sir Isaac Pitman & Sons.

May, R.M., 1994. Biological diversity: differences between land and sea. *Philos. Trans. R. Soc. Lond. B* **343**, 105–111.

Odum, E.P., 1975. Diversity as a function of energy flow. In: van Dobben W.H., Lowe-McConnell R.H. (Eds), *Unifying Concepts in Ecology.* The Hague: Junk, pp 11–14.

Ozenda, P., 2000. *Les végétaux. Organisation et diversité biologique*, 2nd edition. Paris: Dunod.

Paine, R.T., 1966. Food web complexity and species diversity. *Am. Nat.* 100, 65–75.

Pickett, S.T.A. and White, P.S., 1985. *The Ecology of Natural Disturbance and Patch Dynamics.* Orlando, FL: Academic Press.

Resh, V.H., Brown, A.V., Covich, A.P., Gurtz, M.E., Li, H.W., Minshall, G.W., Reice, S.R., Sheldon, AL., Wallace, J.B. and Wissmar, R.C., 1988. The role of disturbance in stream ecology. *J. N. Am. Benthol. Soc.* **7**, 433–455.

Slobodkin, LB., 1968. How to be a predator? *Am. Zool.* **8**, 43–51.

Sousa, W.P., 1984. The role of disturbance in natural communities. *Annu. Rev. Ecol. Syst.* **15**, 353–391.

Taberlet, P., Fumagalli, L., Wust-Saucy, A.G. and Cosson, J.F., 1998. Comparative phylogeography and postglacial colonization routes in Europe. *Mol. Ecol.* **7**, 453–464.

Tonn, W.M., 1990. Climate change and fish communities: a conceptual framework. *Trans. Am. Fish. Soc.* **119**, 337–352.

Townsend, C.R., 1989. The patch dynamics concept of stream community ecology. *J. N. Am. Benthol. Soc.* **8**, 36–50.

Wilson, E.O. and Simberloff, D.S., 1969. Experimental zoogeography on islands. Defaunation and monitoring techniques. *Ecology* **50**, 267–278.

Further Reading

Anonymous, 2000. Biodiversité. L'homme est-il l'ennemi des autres espèces? *La Recherche*, Supplement, no. 333, July/August.

Anonymous, 2000. Bioinformatics for biodiversity. *Science* 289, 2305–2314.

Anonymous, 2001. The human genome. *Nature* **409**, 6822.

Aubertin, C. and Vivien, F.D., 1998. *Les enjeux de la biodiversité*. Economica. Poche Environnement.

Bradbury, I.K., 1998. *The Biosphere*, 2nd edition. Chichester: John Wiley & Sons.

Brown, J.H., 1996. *Biogeography*, 2nd edition. London: McGraw-Hill.

Canton, J.T. and Geller, J.B., 1993. Ecological roulette : the global transport of non-indigenous marine organisms. *Science* **261**, 78–82.

Chapin, F.S., Walker, B.H., Hobbs, R.J., Hooper, D.U., Lawton J.H., Sala, O.E. and Tilman, D., 1997. Biotic control over the functioning of ecosystems. *Science* **277**, 500–504.

Combes, C., 1995. *Interactions durables. Ecologie et évolution du parasitisme*. Paris: Masson.

Cornell, H.V. and Lawton, J.H., 1992. Species interactions. Local and regional processes, and 1 imits to the richness of ecological communities: a theoretical perspective. *J. Anim. Ecol.* **61**, 1–12.

Daszak, P., Cunningham, A.A. and Hyatt, A.D., 2000. Emerging infectious diseases of wildlife – biodiversity and human health. *Science* **287**, 443–449.

Dawkins, R., 1989. *The Selfish Gene*, 2nd edition. Oxford: Oxford University Press.

di Castri, F. and Younès, T., 1990. Fonction de la diversité biologique au sein de l'écosystème. *Acta Oecolog.* **11**, 429–444.

Fontenille, D. and Lochouarn, L., 1999. The complexity of the malaria vectorial system in Africa. *Parasitologia* **41**, 267–271.

Fuhrman, J.A., 1999. Marine viruses and their biogeochemical and ecological effects. *Nature* **399**, 541–548.

French Academy of Sciences, 2000. *Systématique. Ordonner la Diversité du Vivant*. Rapport sur la Science et la Technologie no 11. Paris: Editions Tec&Doc.

Fustec, E., Lefeuvre, J.C. *et al.*, 2000. *Fonctions et valeurs des zones humides*. Paris: Dunod.

Gaston, K.J. and Spicer, J.I., 1998. *Biodiversity. An introduction*. Oxford: Blackwell Science Ltd.

Gaston, K.J., 2000. Global patterns in biodiversity. *Nature* **405**, 220–227.

Golley, F., 1993. *A History of the Ecosystem Concept in Ecology: More Than the Sum of the Parts*. New Haven, CT: Yale University Press.

Gould, J., 1980. *The Panda's Thumb: More Reflections in Natural History*. New York: W.W. Norton & Co.

Gould, J., 1989. *Wonderful Life: The Burgess Shale and the Nature of History*. New York: W.W. Norton & Co.

Grenier, C., 2000. *Conservation contre nature. Les îles Galapagos*. Paris: IRD Editions.

Harvell, C.D., *et al.*, 1999. Emerging marine diseases – climate links and anthropogenic factors. *Science* **285**, 1505–1510.

Hector, A., *et al.*, 1999. Plant diversity and productivity experiments in European grasslands. *Science*, **5442**, 1123–1127.

Heywood, V. and Watson, R. (Eds), 1995. *Global Biodiversity Assessment*. UNEP. Cambridge: Cambridge University Press.

Hildrew, A.G., 1996. Whole river ecology: spatial scale and heterogeneity in the ecology of running waters. *Arch. Hydrobiol*. suppl. 113, Large Rivers **10**, 25–43.

Holling, C.S., 1992. Cross-scale morphology, geometry, and dynamics of ecosystems. *Ecol. Monogr*. **62**, 447–502.

Hooper, D.U. and Vitousek, P.M., 1997. Plant composition and diversity effects on ecosystem processes. *Science* **277**, 1302–1305.

Huston, A..M., 1994. *Biological Diversity. The Coexistence of Species on Changing Landscapes*. Cambridge: Cambridge University Press.

Hutchinson, G.E., 1941. Limnological studies in Connecticut: IV. Mechanism of intermediary metabolism in stratified lakes. *Ecol. Monogr*. **11**, 21–60.

Johnson, K.H., 2000. Trophic-dynamic considerations in relating species diversity to ecosystem resilience. *Biol. Rev. (Cambridge)* **75**(3), 347–376.

Johnson, K.H., Vogt, K.A., Clark, H.J., Schmitz, O.J. and Vogt, D.J., 2000. Biodiversity and the productivity and stability of ecosystems. *Trends Ecol. Evol*. **11**(9), 372–377.

Jones, C.G., Lawton, J.H. and Schchak, M., 1994. Organisms as ecosystem engineers. *Oikos* **69**, 373–386.

Lévêque, C., 1997. *Biodiversity Dynamics and Conservation. The Freshwater Fish of Tropical Africa*. Cambridge: Cambridge University Press.

Lévêque, C., 2001. *Ecologie. De l'écosystéme à la biosphére*. Paris: Dunod,.

Levin, S., 1999. *Fragile Dominion. Complexity and the Commons*. Cambridge, MA: Perseus Books.

Levin, S. (ed.), 2000. *Encyclopedia of Biodiversity*. London: Academic Press.

Loreau, M., Naeem, S., Inchausti, P., Bengtsson, J., Grime, J P., Hector, A., Hooper, D.U., Huston, M.A., Raffaelli, D., Schmid, B., Tilman, D, and Wardle, D.A., 2001. Ecology: biodiversity and ecosystem functioning: Current knowledge and future challenges. *Science* **294**, 804–808.

Mills, L.S., Soulé, M.E. and Doak, D.F., 1993. The keystone species concept in ecology and conservation. *BioScience* **43**, 219–224.

Mooney, H.A., 1995. *Functional Roles of Biodiversity: A Global Perspective*. New York: John Wiley & Sons.

Myers, N. and Knoll, A.H., 2001. The biotic crisis and the future of evolution. *Proc. Natl Acad. Sci. U. S. A.* **98**(10), 5389–5392.

Myers, N., Mittermeier, R.A., Mittermeier, C.G., da Fonseca, G.A.B. and Kent, J., 2000. Biodiversity hotspots for conservation priorities. *Nature* **403**, 853–858.

Naem, S., 1998. Species redundancy and ecosystem reliability. *Cons. Biol*. **12**(1), 39–45.

Pace, N.R., 1997. A molecular view of microbial diversity and the biosphere. *Science* **276**, 734–740.

Parizeau, M.H., 1997 (Ed.). *La biodiversité. Tout conserver ou tout exploiter?* Bruxelles: DeBoeck Université.

Perrings, C., Holling, C.S., Karl-Göran M., Bengt-Owe, J. and Folke, C., 1997. *Biodiversity Loss. Economic and Ecological Issues*. Cambridge: Cambridge University Press.

Powell, J.R., Petrarca, V., della Torre, A., Caccone, A. and Coluzzi, M., 1999. Population structure, speciation, and introgression in the *Anopheles gambiae* complex. *Parassitologia* **41**, 101–115.

Primak, R.B., 1995. *A Primer of Conservation Biology*. Sunderland MA: Sinauer Associates Inc.

Rosenzweig, M.L., 2001. Loss of speciation rate will impoverish future diversity. *Proc. Natl Acad. Sci. U. S. A.* **98**(10), 5404–5410.

Schulze, E.D. and Mooney, H.A. (Eds), 1993. *Biodiversity and Ecosystem Function*. New York: Springer Verlag.

Moon-van der Stay, S.Y., De Wachter, R., Vaulot, D., 2001. Oceanic 18S rDNA sequences from picoplankton reveal unexpected eukaryotic diversity. *Nature* **409**, 607–610.

Steele, J.H. and Schumacher, M., 2000. Ecosystem structure before fishing. *Fish. Res.* **44**, 201–205.

Swanson, T.S., 1995. *The Ectoprocts and Ecology of Biodiversity Decline. The Forces Driving Global Change*. Cambridge: Cambridge University Press.

Tilman, D., Knops, J., Wedin, D., Reich, P., Ritchie, P. and Siemann, E., 1997. The influence of functional diversity and composition on ecosystem processes. *Science* **277**, 1300–1302.

Tilman, D., Wedin, D. and Knops J., 1996. Productivity and sustainability influenced by biodiversity in grassland ecosystems. *Nature* **379**, 718–720.

van der Heijden, M.G.A., Klironomos, J.N., Ursic, M., Moutoglis, P., Streitwoif-Engel, R., Boiler, T., Wiemken, A. and Sanders, I.R., 1998. Mycorrhizal fungal diversity determines plant biodiversity, ecosystem variability and productivity. *Nature* **396**, 69–72.

Vitousek, P.M., Mooney, H.A., Lubchenko, J. and Melillo, J.M., 1997. Human domination of earth's ecosystems. *Science* **277**, 494–499.

Walker, B., Kinzig, A. and Langridge, J., 1999. Plant attribute diversity, resilience, and ecosystem function: the nature and significance of dominant and minor species. *Ecosystems* **2**, 95–113.

Wilson, E.O., 1992. *The Diversity of Life*. Cambridge, MA: Belladone Press.

WRI/IUCN/UNEP, 1992. *Global Conservation Strategy. Guidelines for action to save, study and use Earth's biotic wealth sustainably and equitably*. World Resource Institute, Washington; World Conservation Union, Gland; United Nations Environment Programme, Nairobi.

Index

Biodiversity   Christian Lévêque and Jean-Claude Mounolou
© 2004 John Wiley & Sons, Ltd   ISBN 0 470 84956 8 (Hbk)   ISBN 0 470 84957 6 (pbk)

AAP-0837